Charles Evan Fowler

The coffer-dam Process for Piers

Practical Examples from actual Work

Charles Evan Fowler

The coffer-dam Process for Piers
Practical Examples from actual Work

ISBN/EAN: 9783337106324

Printed in Europe, USA, Canada, Australia, Japan

Cover: Foto ©ninafisch / pixelio.de

More available books at **www.hansebooks.com**

John Wiley & Sons.

THE COFFER-DAM PROCESS FOR PIERS.

FOWLER.

THE
FER-DAM PROCESS FOR PIERS.

PRACTICAL EXAMPLES FROM ACTUAL WORK.

BY

CHARLES EVAN FOWLER,

Member American Society of Civil Engineers,
Bridge Engineer.

"Much of the success of any one in any kind of work, and especially in work subject to the peculiar difficulties of that we are considering, depends upon the spirit in which it is undertaken."— ARTHUR MELLEN WELLINGTON.

FIRST EDITION.

FIRST THOUSAND.

NEW YORK:
JOHN WILEY & SONS.
LONDON: CHAPMAN & HALL, LIMITED.
1898.

Copyright, 1898,
BY
C. E. FOWLER.

74962

ROBERT DRUMMOND,
AND
D. H. RANCK PUB. CO.,
Electrotypers.

ROBERT DRUMMOND,
Printer,
444 PEARL STREET,
NEW YORK.

INTRODUCTION.

THE greater part of foundation work is of an ordinary character. And while difficult foundations have been quite fully treated by engineering writers, ordinary ones have too often been passed over with mere mention, or treated in such a general way that the information proves of little value in actual practice.

Many valuable examples of work of this character have been described in current engineering literature, and it is hoped that by bringing them together a real service will be rendered the profession, as well as much valuable time be saved for considering other and equally important problems.

The history of the coffer-dam process would seem to indicate that engineers of nearly a century ago gave more consideration to the smaller problems than the engineer of to-day, who has apparently passed to the consideration of the larger and of course more interesting ones.

That this is deplorable, is proven by the many cases where money has been wasted in the after effort to make good the mistakes that have become apparent where cheap construction of coffer-dams has been resorted to. The saving in original cost, as between an indefensible method and a defensible one, is often so small as to seem absurd when it has become necessary to make large expenditures to rectify the errors.

Errors of judgment are more easily excusable with regard to foundations than with any other class of construction, but where definite limits can be set, economy will result by keeping as closely as possible within them.

Reference is made in the following pages to the splendid construction of foundations by the Romans, where they could be built outside the water. The Pont du Gard, illustrated in the frontispiece, is the most notable example of this extant. It is interesting also as indicating their knowledge of the better form of piers and methods of arch construction.

Although constructed during the reign of the Emperor Augustus, at the beginning of the Christian era, it is in a remarkable state of preservation, aside from repairs that have been made from time to time.

Probably the earliest recorded examples of the use of coffer-dams which give details of construction are those constructed under the engineers of the Ponts et Chaussées.

Those built under Perronet at the bridge of Orleans were large and extensive, and references made to the pile drivers and the pumps used on the work, serve to illustrate the great amount of attention paid to planning the details of construction.

The same engineer completed the piers of the bridge at Mantes, where the coffer-dams were constructed to enclose both the abutment and the nearest pier within one dam, making the dimensions about 150 feet by 200 feet in the extreme!

Hardly less notable were the coffer-dams at Neuilly, where the interiors were so large, that the excavation did not approach near the inside wall of the dam.

All of these were constructed prior to the year 1775, and the details as shown in the elaborate drawings are of much interest to the engineer engaged on similar works.

The coffer-dams constructed about 1825 by Rennie on the new London bridge were the prototypes of those used at Buda-Pesth, but were elliptical in form. They were designed with as much care, apparently, as any other feature of the bridge, and from the fact that the water was pumped to twenty-nine feet below low water and the work found tight, the details must have been very carefully executed.

However great the amount of care bestowed, there will be cases undoubtedly where the difficulties cannot be foreseen, and it will become necessary to adopt some of the many expedients cited to overcome them ; or they might better be employed from the start, where any suspicion is had that trouble may ensue.

The question as to whether it will be best to use a crib or a sheet-pile coffer-dam will most always be decided by the character of the bottom, the location, and the character of the foundation to be built. It is advisable, whichever type is selected, to make the size large enough, so that the excavation may be completed without approaching too close to the inside wall of the dam, and so that plenty of room may be had for the laying of the foundation courses.

The unit stress adopted for timber construction is believed to be as large as will give good results in the majority of cases, both on account of the possibility of the construction having to undergo more severe usage than is expected, and on account of the grade of timber which is most often made use of for temporary works.

Where it is permissible from the standpoint of true economy, it is believed that steel construction will commend itself for use. In most localities it will not be long until metal construction will be found cheaper than timber for building coffer-dams, and in many places this is already true.

A great mistake is made, in nearly nine cases out of ten, by trying to use old machinery, such as hoisting engines, pumps, and the like, which are ill adapted to the purposes for which they are intended, on account of lack of capacity and only too often on account of having outgrown their usefulness.

The engineer would avoid many unpleasant situations by demanding that a proper outfit be provided, and in the end gain the thanks of the contractor for increased profits.

Extended acquaintance with Portland cement is increasing the use of concrete in construction, and this is a great gain for the engineer, as it is not only superior to much stone that is used, but is better adapted to use in difficult situations. It also lends itself more readily to use for ornamental details in pier construction. That truly ornamental piers are not, however, those with needless and frivolous details, has been clearly set forth in the last article. Simplicity and beauty are near relatives.

The best locations cannot always be chosen for piers, but careful examination will often be the means by which bad locations may be avoided.

The methods for determining the economic division of a given crossing of a river, have not come into general use, probably on account of lack of easy application. The method given is an accurate one and very simple to use, especially if the results are tabulated for a given loading.

TABLE OF CONTENTS.

ARTICLE I.

HISTORICAL DEVELOPMENT.

PAGE

Relation of Foundation to Bridge Design.—Roman and Other Ancient Foundations.—Bridge at Shuster, Persia.—Roman Arch at Trezzo.—Four Ancient Methods for Foundations.—Method of Open Caissons.—Method with Piles and Concrete Capping.—Method of Encaissement.—Method of Coffer-dams.—Cæsar's Bridge over the Rhine.—Pneumatic Caissons and Coffer-dams applicable to Different Cases.—Origin of Coffer-dams and Primitive Types.—The Hutcheson Bridge at Glasgow.—Robert Stevenson's Specifications for Cofferdams on Hutcheson Bridge.—Old Directions for Triple-puddle Coffer-dam in Forty Feet (!) of Tide-water.—W. Tierney Clark's Account of the Great Cofferdams for the Buda-Pesth Suspension Bridge.—Character of Puddle used.—Class of Work to which Coffer-dams should be applied.—Value of Actual Examples.. 1

ARTICLE II.

CONSTRUCTION AND PRACTICE. CRIB COFFER-DAMS.

Definition of Coffer-dam.—Simple Clay Bank.—Drag Scraper for removing Soft Bottom.—Excavating Spoon.—Larger Dredges mentioned.—Crib and Embankment used on Chanoine Dams on Great Kanawah River.—Improvised Nasmyth Sheet-pile Hammer.—Failure on Ohio River because of Porous Bottom. —Crib Coffer-dam with Puddle Chamber, C., B. & Q. R. R.—Cribs without Puddle Chambers, Can. Pac. Ry.—Cribs of Old Plank, Santa Fé Ry.—Crib for Arkansas River, St. L. and S. F. Ry.—Sheet Piles used on Santa Fé.—Sheet Piles used on Union Pacific Ry.—Coffer-dam on Grillage, Union Pacific Ry.—Circular Coffer-dam of Staves at Fort Madison, Ia.—Circular Coffer-dam Failure at Walnut St., Phila.—Probable Cause of Failure.—Form of Construction to adopt.—Use of Puddle.—Cutwaters.—True Economy of Construction.. 13

ARTICLE III.

CONSTRUCTION AND PRACTICE. CRIBS AND CANVAS.

Stopping Leaks.—Canvas Bulkhead at Keokuk, Iowa.—Canvas Funnel for Springs. —Anchoring Cribs and Crib Coffer-dam at St. Louis.—Timber Casings cov-

vii

ered with Canvas, Melbourne.—Strength of Water-soaked Timber.—Polygonal Crib for Harlem Ship Canal Pivot Pier.—Polygonal Crib for Arthur Kill Bridge.—Octagonal Crib, Coteau Bridge 28

ARTICLE IV.

PILE DRIVING AND SHEET PILES.

Historical Forms of Pile Drivers.—Simple Sheet-pile Driver.—Large Pile-driving Derricks.—Machinery for Pile Driving.—Cost of Outfits.—Nasmyth Hammers of Various Types.—Loads on Guide and Foundation Piles.—Pulling Piles and sawing off under Water.—Forms of Sheet Piles.—Wakefield Sheet Piling.— Shoes for Sheet Piling... 40

ARTICLE V.

CONSTRUCTION WITH SHEET PILES.

Water and Puddle Pressure.—Calculation of Sheet Piling.—Size of Wales and Struts.—Width of Puddle Chambers.—Guide Piles and Guides.—Ann Arbor Sheet-pile and Puddle Coffer-dam, M. C. Ry.—Failure with Sheet Piles at Arthur Kill Bridge.—Successful Method adopted.—Sewer Coffer-dam for Boston Sewerage System.—Wakefield Sheet Piling.—Harper's Ferry Cofferdam.—Momence, Ill., Coffer-dam, C. & E. I. Ry.—Sheet Piling for Charlestown Bridge Piers.—Polygonal Sheet-pile Reservoir Coffer-dam at Fort Monroe, Va.. 54

ARTICLE VI.

CONSTRUCTION WITH SHEET PILES.

Combinations of Various Forms of Sheet Piles.—Sheet-pile and Puddle Cofferdam, Walnut Street Bridge, Chattanooga.—Framing of Cumberland, Md., Coffer-dam.—Sandy Lake Coffer-dam and Pile-driving Plant.—Driving Sheet Piles with Water Jet.—Use of Sheet Piling on Foundations of Main Street Bridge, Little Rock.—Concrete Piers at Little Rock.—Removal of Old Pier at Stettin, Germany.—Removal and Repair of Pier in Coosa River, Alabama.— Floating Coffer-dam for P. & R. R. R. Bridge over the Schuylkill.—Use of Six-inch Sheet Piles at St. Helier, Jersey.—Stock Rammer to stop Leaks.—Single-pile Coffer-dams, Putney Bridge.—Twelve-inch Sheet Piling, Victoria Docks.— Tongue and Groove Sheet Piling, Topeka, Kansas.—Use of Dredging Pump at Topeka.. 66

ARTICLE VII.

METAL CONSTRUCTION.

Thin Steel Shells.—Hawkesbury Oblong Metal Piers.—Vertical and Inclined Cutting Edges.—Water-tight Construction.—Pivot Pier of Clustered Cylinders.—

Double-cylinder Pier.—Russian Ornamental Cylinder Piers.—Lighthouse Cylinders.—Calculation of Thin Metal Cylinders.—Forth Bridge Metal Cofferdams.—Forth Bridge Circular Granite Piers.—Combined Metal Coffer-dam and Pier Base.—Metal Sheet Piles 80

ARTICLE VIII.

PUMPING AND DREDGING.

Amount of Pumping indicates Success.—Bascule for Pumping.—Chapelet for Pumping.—Bucket Wheel used at Neuilly.—Box Lift Pump.—Metal Lift Pump.—Diaphragm Pump.—Steam Siphons.—Van Duzen Jet.—Lansdell Siphon.—Pulsometer Steam Pump.—Maslin Automatic Vacuum Pump.—Comparative Efficiency of Centrifugal and Reciprocating Pumps.—Tests of Centrifugal Pumps.—Direct-Connected Engine and Centrifugal Pumps.—Use of Electric Power.—Suction-pipe Details.—Type and Capacity of Pump.—Methods of Priming.—Double-suction Pumps.—Dredging Pumps.—Clam shell and Grapple Dredges.—Sand Diggers and Elevator Dredges.—Dipper Dredges.—Cost of Dredging 92

ARTICLE IX.

THE FOUNDATION.

Character of Foundation.—Kind of Bottom.—Soft Bottom.—Pile Foundation.—Soft Material overlying Hard Bottom.—Clean Smooth Rock.—Sloping Rock.—Rough Rock.—Concrete Levelling Course.—Concreting under Water.—Monolithic Concrete Piers.—Concrete Piers at Red River.—Monolithic Concrete on Illinois and Mississippi Canal.—Requirements for Good Concrete.—Composition of Concrete.—Contractor's Plant.—Cableways 106

ARTICLE X.

LOCATION AND DESIGN OF PIERS.

Location at Fixed Site.—Location at New Site.—Government Requirements.—Examination of Site.—Test-boring Apparatus.—Mississippi River Commission Boring Device.—Economical Length of Spans.—Ottewell's Formula for Economic Span.—Morison's Design for Piers.—Omaha Union Pacific Piers.—Russian Piers.—Obstruction caused by Piers.—Cresy's Experiments on the Obstruction caused by Piers.—Correlation of Theoretical Form and Architectural Design 120

TABLE OF COFFER-DAMS.—

No.	Page	River and Location.	Current.	Water Head.	Character of Bottom.
1	5	River 200 feet wide, Ohio	None.	12' +	Cemented gravel.
2	6	Clyde at Glasgow	Slight.	9' +	Gravel, sand, mud.
3	8	Estuary or Harbor	Tide.	40'	Sand & gravel over clay.
4	9	Danube at Buda Pesth	Swift.	54' ±	Gravel over clay.
5	14	Kanawah near mouth	Swift.	34' −	Gravel over hardpan.
6	15	Ohio near head	Moderate.	20' +	Gravel.
7	17	Western part United States	Moderate.	6' +	Soft.
8	18	St. Lawrence, lower river	Swift.	20' +	Rock.
9	18	Arnprior Bridge	Swift.	21' +	Rock.
10	18	New Mexico, underflow	None.	15' +	Sand.
11	20	Arkansas at Tulsa	Moderate.	7' +	Gravel over rock.
12	20	Western part United States	Moderate.	6' +	Soft.
13	20	Republican in Kansas	Moderate.	6' +	Sandy.
14	20	Western part United States	Moderate.	7' +	Gravel over soapstone.
15	20	Western part United States	Moderate.	6' +	Rock.
16	20	Payette and Weiser, Union Pacific.	Moderate.	6'	Soft.
17	20	Mississippi, Fort Madison	Swift.	19'	Soft.
18	24	Schuylkill near Philadelphia, Pa...	Moderate.	Deep.	Mud over rock.
19	30	U. S. Canal, Keokuk	None.	12' +	Rock.
20	32	Mississippi, St. Louis	Swift.	22'	Rock.
21	33	Queen's Bridge	Swift.	15'	Rock.
22	36	Harlem Ship Canal	Moderate.	25'	Rock.
23	38	Arthur Kill Bridge	Tide.	28'	Clay over rock.
24	39	Coteau Bridge, C. Pac. Ry.	Moderate.	28'	Rock.
25	59	Ann Arbor, Mich., M. C. Ry.	Moderate.	6' +	Gravel.
26	60	Arthur Kill Bridge	Tide.	30' −	Mud and clay.
27	60	Boston Harbor, sewer	Tide.	10'	Sand and gravel.
28	62	Illinois River, La Grange	Moderate.	7'	Sand and mud.
29	62	Kankakee at Momence	Moderate.	6' +	Rock.
30	63	Potomac at Harper's Ferry	Swift.	6' +	Rock.
31	63	Charlestown Bridge, Boston	Tide.	6' +	Soft.
32	64	Fort Monroe sewer	None.	20'	Soft.
33	66	Tennessee at Chattanooga	Swift.	8' +	Gravel over rock.
34	67	Cumberland, Md	Moderate.	10' +	Sand over hardpan.
35	67	Mississippi. Sandy Lake	Swift.	8' +	Sand.
36	70	Arkansas, Little Rock	Moderate.	6' +	Sand.
37	72	Parnitz, Stettin, Germany	Moderate.	25' +	Clay.
38	74	Coosa, Gadsden, Ala	Moderate	10' +	Gravel over rock.
39	74	Schuylkill, P. & R. R. R.	Swift.	8' +	Rock.
40	77	St. Heller Bridge, Jersey, Eng.	Tide.	13' +	Earth over rock.
41	77	Thames at Putney	Moderate.	Deep.	Mud.
42	77	Victoria (B. C.) Docks	Tide.	35'	Rock.
43	78	Kaw at Topeka	Swift.	6' +	Sand.
44	86	Firth of Forth	Tide.	15' +	Rock.

x

SYNOPSIS OF EXAMPLES.

Form of Construction.	Inside Dimensions.	Kind of Puddle.	Thickness Puddle.	Remarks.	Page	No.
Earth bank.	10' × 60'?	Clay and gravel.	5' +	No leaks.	5	1
Sheet piles.	20' × 58'?	Clay.	3'		6	2
Sheet piles.	Large.	Clay, sand & gravel.	3-6'	Typical.	8	3
Sheet piles.	72' × 136' +	Clay and gravel.	2-5'	Difficult.	9	4
Earth bank.	90' × 330'	Clay and gravel.	19' +		14	5
Earth bank.?	200' × 600'	Clay and gravel.		Failed.	15	6
Crib.	Medium.	Clay.	3' +		17	7
Crib, single.	24' × 43'	Concrete inside.			18	8
Crib, single.	16' × 34'	Concrete inside.			18	9
Crib, single.	17' × 43'	Clay outside.		Special.	18	10
Crib, single.	Medium.	Clay outside.			20	11
Sheet piles.	Medium.			Typical.	20	12
Sheet piles.	Medium.	Clay outside.			20	13
Sheet piles.	Medium.	Clay outside.			20	14
Sheet piles.	Medium.	Clay.	Equal depth.		20	15
Box or crib.	12' × 36'	None.		On grillage.	20	16
Staves.	36' diam.	None.		On grillage.	20	17
Sheet piles.	80' diam.	None.		Failed.	24	18
Canvas on plank.	80' long.	Rotten manure.		Bulkhead.	30	19
Crib, double.	28' × 64'	Clay.	3' 0"	Canvas used	32	20
Box and canvas.	Square.	Clay outside.		Movable.	33	21
Polygon crib.	47' diam.	Clay.	4' 6"		36	22
Polygon crib.	44' diam.	Clay and gravel.	5' 0"		38	23
Crib, single.	34' diam.	Concrete inside.			39	24
Sheet piles.	13' × 44'	Clay and gravel.	2' 8"		59	25
Sheet piles.	Large.	None.		Two trials.	60	26
Sheet piles.	12' wide.	Clay.	6' to 8'		60	27
Sheet piles.	Medium.	None.			62	28
Sheet piles.	Medium.	Gravel.		Two trials.	62	29
Sheet piles.	Medium.	Gravelly clay.			63	30
Sheet piles.	18' 6" × 119'	Concrete inside.			63	31
Sheet piles.	44' diam.	Sand and concrete.	7' +		64	32
Sheet piles.	Large.	Clay.	9' 0"		66	33
Sheet piles.	15' × 50'	None.			67	34
Sheet piles.	829' long.	Clay.	8' ±		67	35
Sheet piles.	16' × 38'	Earth outside.			70	36
Sheet piles.	23' × 55' ±	Clay.	2' to 4'	Removal.	72	37
Sheet piles.	28' × 28' ±	Clay.	12' +	Removal.	74	38
Sheet piles.	16' × 42'	Clay and gravel.	8' +	Movable.	74	39
Sheet piles.	Medium.	Clay outside.			77	40
Sheet piles.	Medium.	None.			77	41
Sheet piles.	500' long.	Clay.	2-7'		77	42
Sheet piles.	18' × 55'	Clay outside.			78	43
Metal.	60' diam.	Concrete seal.			86	44

LIST OF ILLUSTRATIONS.

NUMBER		PAGE
	The Pont du Gard, Nîmes, France....................................	*Frontispiece.*
1.	Bridge at Shuster, Persia, over the River Karun..................	2
2.	Bridge over the Adda at Trezzo Milanese..........................	3
3.	Cæsar's Bridge over the Rhine....................................	4
4.	A Primitive Solution. (*Earth-bank Coffer-dam.*)..................	6
5.	Coffer-dam in Tide-water. (*Sheet Piles and Puddle.*).............	8
6.	Buda-Pesth Suspension Bridge. (*Puddle Coffer-dam.*)..............	9
7.	Buda-Pesth Suspension Bridge, Plan of Coffer-dam No. 3............	11
8.	*Scraper Dredge.* (For Drag Dredging, C. & M. V. Ry.).............	14
9.	Coffer-dam at Dam No. 11, Gt. Kanawah River. (*Earth and Crib.*)..	15
10.	*Crib* Coffer-dam, C., B. & Q. R. R. (*With Puddle Chamber.*)....	16
11.	St. Lawrence River Bridge, C. P. Ry. (*Crib and Coffer-dam*).....	17
12.	Arnprior Bridge, C. P. Ry. (*Crib and Coffer-dam*)...............	18, 19
13.	*Crib* Coffer-dam, A., T. & S. F. Ry. (*No Puddle Chamber.*).....	21
14.	*Coffer-dam on Grillage,* Payette and Weiser Rivers, U. P........	22, 23
15.	*Coffer-dam on Grillage,* Fort Madison Bridge, A., T. & S. F. Ry.	24
16.	A Crib *Coffer-dam* after a Flood. (*Showing Plant.*)............	25
17.	Apparatus used to force Clay into Crevice of Rock. (*Leak.*).....	29
18.	Details of *Canvas and Plank Bulkhead,* Keokuk, Ia...............	31
19.	Inside View of *Bulkhead,* Lock pumped Dry, Keokuk, Ia...........	34
20.	*Canvas Funnel* for closing *Leaks.* (Springs.)..................	35
21.	*Cribs* for anchoring St. Louis Coffer-dam. (*Crib and Puddle.*).	36
22.	Polygonal (*Crib*) *Coffer-dam.* Harlem Ship Canal Bridge........	38
23.	Details Coffer-dam, Arthur Kill Bridge. (*Crib and Puddle.*).....	37
24.	Coffer dam for Pivot Pier, Coteau Bridge. (*Crib.*)..............	38
25.	Perronet's Pile Driver. (Historical; Man Power.).................	41
26.	Perronet's Bull-wheel Pile Driver. (Historical; Horse Power.)....	41
27.	Sheet-pile Driver. (Hand-power Derrick.).........................	41
28.	Pile-driver Derrick for Use on a Scow............................	42
29.	Lidgerwood Pile-driving Derrick..................................	43
30.	Hammer with Nippers. (For Horse Power.)..........................	43
31.	Pile-driving Scow, New York State Canals. (Steam.)...............	44
32.	Warrington-Nasmyth Steam Pile Hammer.............................	45
33.	Warrington-Nasmyth Hammer, Fair Haven Bridge.....................	46
34.	Cram-Nasmyth Steam Pile Hammer...................................	47
35.	Machine for sawing off Piles under Water.........................	48
36.	Pile-pulling Lever. (Hand Power.)................................	49
37.	Pile-pulling Scow, New York State Canals. (Steam.)...............	50

LIST OF ILLUSTRATIONS.

NUMBER	PAGE
38. Sheet Piles and Sheet-pile Details	51
39. Charlestown Bridge. Driving Wakefield Sheet Piling	52
40. Arrangement and Diagrams of Sizes for Sheet-pile Coffer-dams	55
41. Sheet-pile Guides and Clamps	57
42. Coffer-dam for Ann Arbor Bridge, M. C. Ry. (Sheet Piles and Puddle.)	58
43. Sewer Coffer-dam, Boston Sewerage System. (Sheet Piles and Puddle.)	59
44. Wakefield Sheet Piling. (Details.)	61
45. Type of Momence and Harper's Ferry Coffer-dams. (Sheet Piling.)	62
46. Coffer-dam on Charlestown Bridge. (Sheet Piling.)	63
47. Resevoir Coffer-dam, Fort Monroe, Va. (Sheet Piling.)	65
48. Compound Sheet Pile	67
49. Chattanooga Bridge, Bed-rock Pier No. 3	68
50. Framework of Coffer-dam, Cumberland, Md. (Sheet Piling)	69
51. Sandy Lake Coffer-dam. (Sheet Piling.)	70
52. Coffer-dam and Concrete Pier, Little Rock, Ark. (Sheet Piling.)	71
53. Removal of Masonry Pier at Stettin, Germany. (Sheet Piling.)	73
54. Coosa River Coffer-dam. (Sheet Piling.)	75
55. Stock Rammer. (For packing Clay to stop Leaks.)	77
56. Topeka Bridge Coffer-dam. (Sheet Piling.)	78
57. Hawkesbury Bridge, Caisson No. 6. (Metal Shell.)	81
58. Group of Cylinders for Pivot Pier. (Metal Shells.)	82
59. Pier of Two Cylinders, Victoria Bridge. (Metal Shells.)	83
60. Circular Saw for cutting off piles under Water	84
61. Cylinder-pier Bridge, Riga-Orel R. R., Russia. (Metal Shells.)	85
62. Cylinder Piers, with Diaphragm. (Metal Shells.)	86
63. Circular Granite Pier, Forth Bridge	87
64. Forth Bridge. (Metal Coffer-dam)	88
65. Forth Bridge. (Circular Granite Pier and Metal Coffer-dam)	90
66. Old Bascule Pump. (Hand Power.)	93
67. Old Chapelet, Side Eelevation. (Water-power Pump.)	94
68. Old Chapelet, End Elevation. (Water-power Pump.)	94
69. Hand Pump, Soldered Joints	95
70. Hand Pump, Screw Joints	95
71. Diaphragm Pump. (Hand Power.)	95
72. Van Duzen Jet Pump. (Steam Power.)	96
73. Lansdell's Siphon Pump. (Steam Power.)	96
74. Pulsometer Steam Pump	97
75. Section of Pulsometer	97
76. Centrifugal Pump, directly connected to Engine	98
77. Suction Details for Pumps	99
78. Centrifugal Pump, Double Suction	100
79. Dredging Pump	100
80. Dredging-pump Piston	101
81. Lancaster Grapple. (Derrick Dredge.)	102
82. Sand Digger. (Light Elevator Dredge.)	103
83. Osgood Dipper Dredge, New York State Canals	104
84. Osgood Dipper Dredge, Details, New York State Canals	105
85. Metal Tube for Concreting	107
86. Metal Bucket for Concreting	108

LIST OF ILLUSTRATIONS.

NUMBER		PAGE
87.	Concrete Piers, Red River Bridge	109
88.	Concrete Forms, Red River Bridge	110
89.	Concrete Forms, Illinois and Michigan Canal	111
90.	Stone Crusher and Concrete Mixer, I. and M. Canal	112
91.	Double-drum Guy Derrick, Am. Hoist & Derrick Co.	113
92.	Single-drum Horse Power, Con. Plant Mfg. Co.	114
93.	Double-drum Hoist Engine, Lidgerwood Mfg. Co.	114
94.	Crocker-Wheeler Electric Hoist	115
95.	Lidgerwood Cableway Carriage and Skip	116
96.	Lidgerwood Cableway at Coosa Dam. (Span 1012 Feet)	118
97.	Hand Drill and Swab	121
98.	Steam-power Well Driller	122
99.	Test-boring Apparatus, Mississippi River Commission	123
100.	Clamp and Maul. (Test Boring.)	124
101.	Pier of Omaha Bridge, Union Pacific System	126
102.	Russian Pier, Russian State Railways	127
103.	Cresy's Experiments on the Form of Piers	128
104.	Cresy's Experiments on the Form of Piers	130

ARTICLE I.

THE COFFER-DAM PROCESS FOR PIERS.

HISTORICAL DEVELOPMENT.

THE continued increase in the weight of our bridge superstructures and of the loads they have to carry has led to increased care, to a very gratifying degree, in the preparation of the foundations for bridge piers and abutments.

An old authority very truly states "The most refined elegance of taste as applied in the architectural embellishment of the structure; the most scientific arrangement of the spans and disposition generally of the superior parts of the work; and the most judicious and workmanlike selection and subsequent combination of the whole materials composing the edifice, are evidently secondary to the grand object of producing certain firm and solid bases whereon to carry up to any required height the various pedestals of support for the spans of the bridge."

There is every reason to believe, from the bridges of the Romans still extant and of those of ancient and mediæval times of which there are remains or records, that the foundations were carefully considered.

The most ancient form was likely begun by dumping in loose stones until the surface of the water was reached and the masonry could then be commenced without the necessity for any method of excluding the water. The oldest civilizations were in tropical or semi-tropical countries where the streams are dry beds for many months in the year and suitable foundations were easily made without water to contend with. Where the bottom of the stream was rock, the engineering could be very little improved upon to-day, and even where there was shallow water on rock bottom, the piers were well founded in the shallowest places, the bridge often winding across the stream in serpentine form, similar to the bridge over the river Karun, at Shuster, Persia. Fig. 1.

The arch was developed to such an extent by the Romans, and the spans were increased to a length which rendered the construction of piers in the water unnecessary for short bridges, the abutments or skewbacks being without the stream on either bank.

The difficulty of founding piers in midstream was doubtless the controlling cause for the larger spans, such as the one built at Trezzo, over the river Adda, by order of the Duke of Milan, sometime prior to the year 1390. The span at low water was 251 feet, the single arch being of granite in two

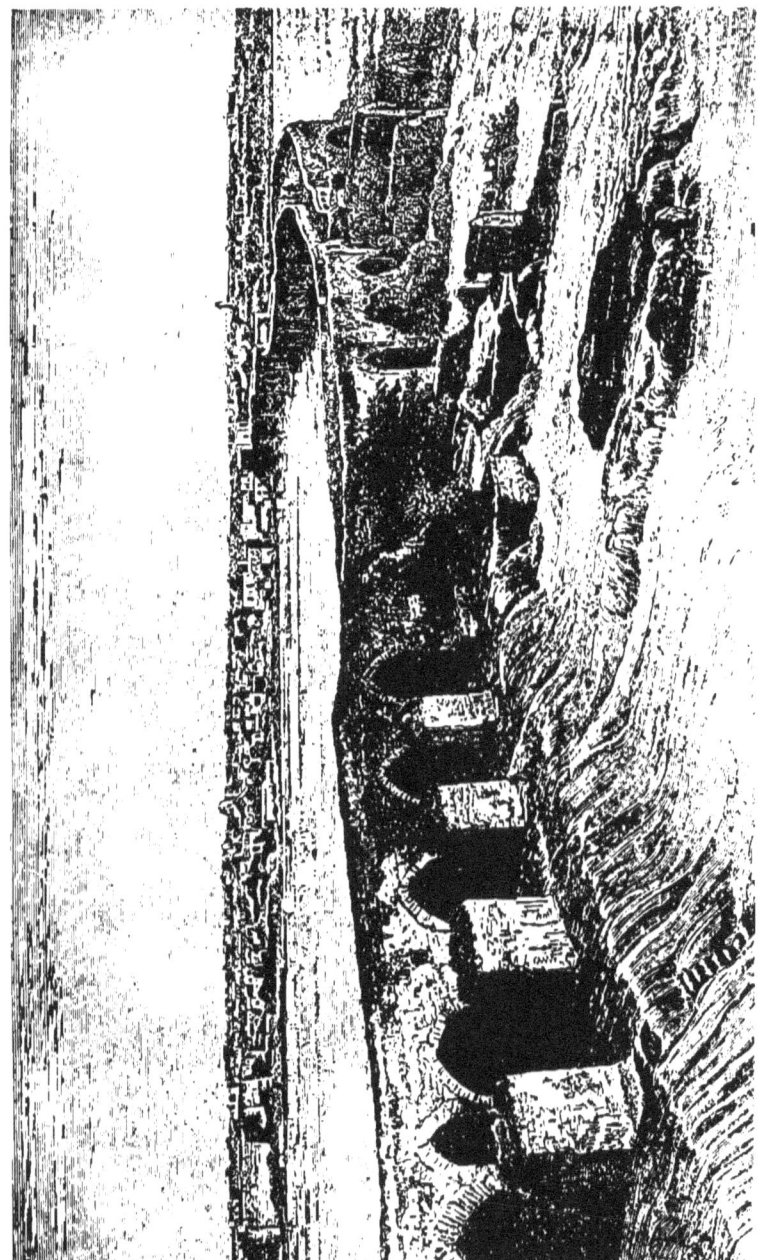

FIG. 1.—BRIDGE AT SHUSTER, PERSIA, OVER THE RIVER KARUN.

courses. The placing of a middle support was doubtless found to be impracticable and caused the design of an arch which has never been equaled or eclipsed. Fig. 2.

The construction of roads has ever been the harbinger of civilization, and with the spread of civilization came a demand for the improvement of means of communication. The engineer was called upon to construct better and greater bridges in a permanent manner, which led to the origin and development of the four methods for founding in water that were used in

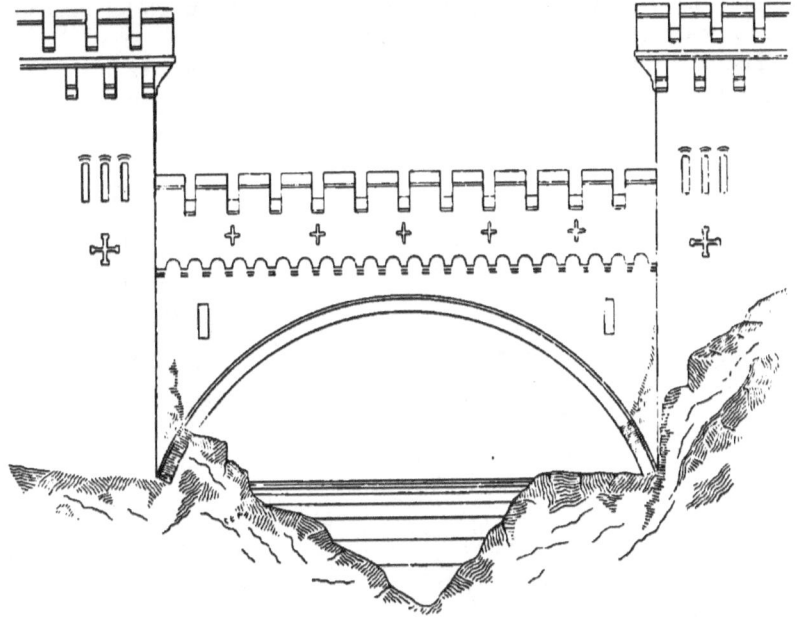

FIG. 2.—BRIDGE OVER THE ADDA, AT TREZZO, MILANESE, A PROBABLE RESTORATION. THE SHADED PORTION OF ARCH RINGS IS ALL THAT REMAINS.

olden times. These may be classified as, first, the method with open caissons; second, the use of piles with a capping of coarse concrete about the tops; third, the use of piles after the manner of the French encaissement; and fourth, the use of coffer-dams. A fifth method might be added, in which the bridge was built on dry land adjacent to the stream, and the river diverted to a new channel afterwards excavated under the completed structure. This is, however, an avoidance rather than a solution, unless the river is to be diverted in the course of its improvement.

The first method, as described in old treatises or accounts, consisted of little more than baskets formed of branches of trees, weighted with stone to sink them, and after sinking filled with loose stone to near low water level, where the masonry could be commenced. These baskets were similar in construction to the mattresses used in the bank revetment of the Mississippi or the bamboo casings used by the Japanese to hold stones in place on bank protection.

An improvement was effected by using in place of baskets, boxes or small open caissons which were sunk and filled in the same manner, several being used for one pier. This was the method used at Blackfriars bridge and also at Westminister bridge, over the Thames, and has been much used in recent times, the caisson being built large and strong enough for the entire pier, which is built up as the caisson sinks.

The second method consisted of driving piles over the area of the foundation until the heads were below low water level, and spaced at distances

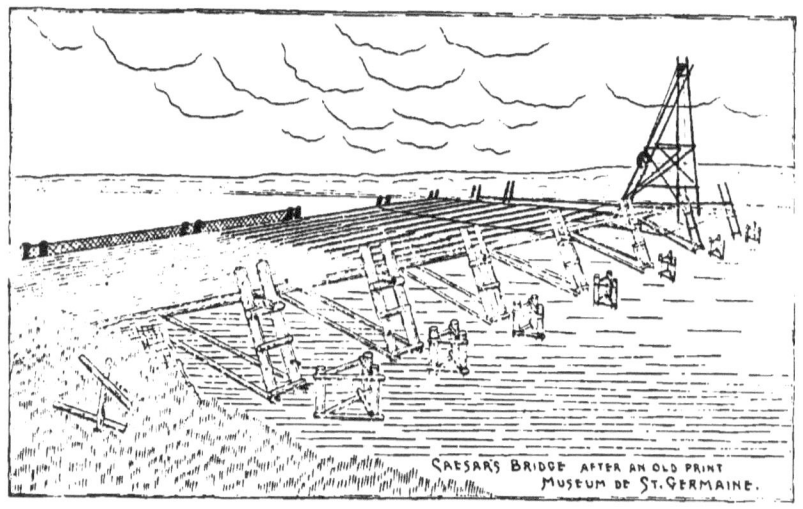

FIG. 3.—TEN DAYS TO CONSTRUCT; LENGTH ABOUT A QUARTER MILE; DEPTH WATER 16′; WIDTH, 25′; BEAM, 2′ THICK; ABOUT 50 PIERS.

apart as required by the nature of the bottom, similar to the methods in vogue to-day. The heads of the piles were not driven to the same level, however, and were incased in a form of coarse concrete such as was used by the Romans, but what is now called beton. This was leveled up and on it was laid the stone for the footing course of the pier.

The third method of encaissement was probably an improvement of the dumping in of loose stone on which to place the pier, and consisted in inclosing the space for the pier with sheet piling, after which the loose material was removed from the bottom as much as possible and the stone dumped inside until nearly up to low water, at which time the pier could be begun.

These last two methods doubtless met with much favor owing to the familiarity with pile driving, in which the Romans especially were proficient. Cæsar's bridge over the Rhine was built entirely on piles, and in a view of it after the old print in the Museum de St. Germaine, is pictured a pile driver on a float in position for driving. Fig. 3.

This third method was the early type of the crib which has been such a factor in the building of the earlier foundations over our American rivers. Crossed timbers laid up crib fashion with rectangular openings or cells between the timbers were sunk and filled with broken stone on which to build the pier.

These methods were all deficient in affording no means of seeing or making a careful examination of the bottom on which the foundation was to be placed, and with the advent of more permanent structures of greater magnitude the coffer-dam came into use. This allowed the bottom to be freed from water and after a careful examination and preparation of the foundation, the work could proceed in the dry until above water level.

The pneumatic caisson is now in general use for all foundations that must go to any considerable depth below the water and has even been used in some instances where the depth was slight, but where for various reasons it was deemed expedient to use compressed air caissons. Recent expressions from some engineers of high standing would indicate that they do not consider it good practice to use coffer-dams in any case, one making the statement that he had not used a coffer-dam for thirty years, while another seemed to think it a matter to be left to the pleasure of the contractor. That the use of this method has gotten into disfavor seems to be beyond question and it will be the purpose of the succeeding pages to learn to some extent why this is so, but mainly to show from successful examples how to proceed, that success instead of failure may result. Any attempt to account for the origin of the coffer-dam process would be futile, inasmuch as the savage, wishing to free a space from water, doubtless banked up earth about the area and scooping out the water with his hands, laid the ground bare for inspection. From so simple a beginning, the first method likely to occur to a mind capable of reasoning, can readily be imagined the course of development of coffer-dams.

The most simple form in use at the present time, where the water is quiet, is shown in the Fig. 4, and consists principally of a bank of earth which is made thick enough to be nearly or quite impervious to water, the earth being prevented from caving into the excavation by piles supporting a timber casing. Some of the recorded examples of the early use of this process are of interest in illustrating the care which was bestowed upon their construction in important works and will call attention to that incessant care which is necessary to success in any work of this character.

Robert Stevenson, the great English engineer, thought it not beneath his dignity to give full instructions as to the construction of the coffer-dams for the Hutcheson bridge over the Clyde at Glasgow. The bridge consisted of five arch spans, the total length between the abutments being 404 feet and the width 38 feet. The four piers were from 11 to 12 feet in thickness, being designed to take up the arch thrust, and 48 feet in length at the foot-

FIG. 4.—A PRIMITIVE SOLUTION.

ing. The specifications written at Edinburgh in April, 1828, are so explicit that they will be quoted in full on this point: "It having been ascertained by boring and mining that the subsoils of the bed of the river consist of gravel, sand and mud to the depth of 27 feet and upwards, it becomes necessary to prepare foundations of pile work for the bridge; and, therefore, to insure the proper and safe execution of the works, coffer-dams are to be constructed around each of the foundation pits of the two abutments and four piers of such dimensions as to afford ample space for driving piles, fixing wale pieces, laying platforms, pumping water, and setting the masonry; and likewise for the construction of an inner or double coffer-dam should this ultimately be found necessary. The framework of the coffer-dams is to consist of not less than two rows of standard or gauge and sheeting piles, kept at not less than three feet apart for the thickness of a puddle wall or dyke, which space is to be dredged to a depth of not less than nine feet under the level of the summer watermark above described, before the sheeting piles are driven. The gauge or standard piles are to measure not less than 24 feet in length and 10 inches square. They are to be placed three yards apart and driven perpendicularly into the bed of the river to the depth of sixteen feet under the level of the summer watermark, thereby leaving eight feet of their length above that mark. Runners or walepieces of timber nine inches square are then to be fitted on both sides of each row of gauge piles, to which they are to be fixed with two screw bolts of not less than one inch in diameter, passing through each of the gauge piles. One set of these inside and outside walepieces is to be placed at or below the level of summer watermark, and the other set within one foot of the top of each row of said piles, the whole to be fixed with screw bolts in the manner above described. The walepieces are to be four and one-half inches apart in order to receive and guide the sheeting piles. This is to be effected by notching the walepieces into the gauge piles. The sheeting piles are to be 21 feet in length, 4½ inches in thickness, and not exceeding 9 inches in breadth. They are to be closely driven, edge to edge, along the space left between the walings, and each compartment of the sheeting between the gauge piles is to be tightened with a key pile. The coffer-dam frames are to be properly connected with stretchers and braces before commencing the interior excavation. Each coffer-dam is to be provided with a draw-sluice, fourteen inches square in the void, with a corresponding conduit passing through the puddle dyke at the level of summer watermark. To render the coffer-dams water tight the whole excavated space between the two rows of piling is to be carefully cleared of gravel, sand or other matters, to the specified depth, and clay well punned or puddled is then to be filled in and carried up to the level of the top of the sheeting piles. But if it shall, notwithstanding, be found that the single tiers of coffer-dam do not keep the foundation pits sufficiently free of water for building operations, the water must either be pumped out and kept perfectly under by steam or other power, or else excluded by the construction of a second tier of coffer-dam similar in construction to the first. For the foundation pits of the two abutment piers on either side of the river it is not expected that more will be required on the landward side for keeping up the stuff than a single row of gauge and sheeting piles; but if the engineer shall find other works necessary upon opening the ground they must be executed by the contractor and shall be paid for agreeably to the contract schedule of prices for the regulation of extra and short works. The stuff within the coffer-dams is to be excavated to the depth of ten feet under the level of summer watermark for each of the piers and eight feet for each of the abutments."

The present practice of leaving all this to a contractor, whose idea is too often to sacrifice everything to cheapness, appears in very unfavorable contrast to this careful description.

An article on founding by means of coffer-dams, published in 1843, gives directions for placing a coffer-dam in forty feet of tide water; and

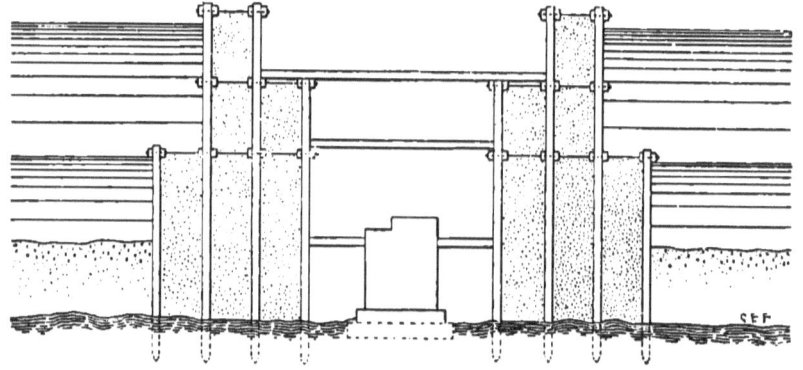

FIG. 5.—COFFER-DAM IN TIDEWATER.

although the engineer of to-day would use some other method for such a depth, an illustration, Fig. 5, and short description of it are given, as ideas may be gained for application to ordinary works.

The water was assumed at ten feet deep for low tide, twenty-eight feet at high tide, with twelve feet of sand and gravel to be removed to expose the clay on which the pier was to rest. Four rows of piles were to be driven around the area, the outer row to within one foot of low water, the two rows in the middle to within three feet of high water, the inner row to eleven feet above low water, and all to be down five feet into the clay. The outer row of piles to be six by twelve inches, the two rows in the middle twelve by twelve inches, and the inner row eight by twelve inches; all driven close together and to have walingpieces, braces and brace rods as shown in cross section. The rows to be six feet apart and to be filled in between with a puddle of clay mixed with sand and gravel.

The report of W. Tierney Clark, the engineer of the Buda Pesth suspension bridge, gives an account of what are probably the largest bridge coffer-dams ever constructed. Some other method would now be used for such a location, but this fact will not detract from the lessons that may be drawn from them.

The Danube was crossed at Buda Pesth previous to the year 1837 by means of a bridge of boats which had to be taken up during the winter and the passage made by ferry or on the ice, so that for six months of each year there was great risk in crossing and frequent loss of life. The building of a permanent bridge was brought about through the efforts of the Count Szechenyi, who, as a member of a committee, proceeded to England in 1832 and after a careful investigation of existing works decided upon the construction of a suspension bridge. The greatest question for solution was the founding of the two towers in a river like the Danube, where the ice

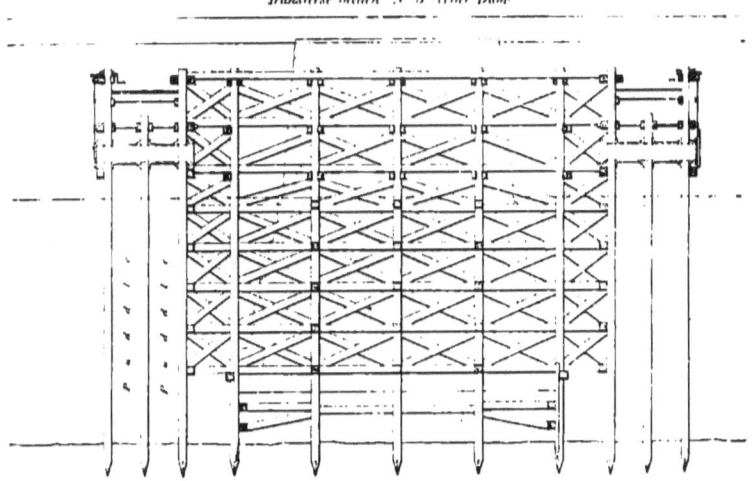

FIG. 6.—BUDA PESTH SUSPENSION BRIDGE.

throughout the long winter wrought havoc with everything in reach. The ice in the river in February, 1838, was from six to ten feet thick near the site of the proposed piers. On March 9 a movement occurred across the whole river and for a length of 350 yards, the whole moving in a solid mass. On March 13 it moved again 400 yards and three hours later a general breaking began. The ice piled up on the shoals causing a sudden rise to twenty-nine feet five inches above zero, and while it was at this height for only a few hours, it is recorded that a great part of Buda and two-thirds of Pesth were destroyed and many lives lost.

The extraordinary design of the coffer-dams can the more readily be understood after this description, it being doubted by many persons at that time whether piers could be placed in the river by any means. Fig. 6.

The drawings reproduced are of coffer-dam No. 3 which was about 72 feet in width and about 136 feet in length inside the puddle walls, there being two puddle chambers, each five feet in width. From a point about thirteen feet above the clay on which the tower was to rest, was an inside wall of sheet piling, this space being nearly filled after excavating, with a bed of concrete. The piling of each row, from forty to eighty feet in length, was all carefully sized to fifteen inches square, shod with iron and driven close together, penetrating twenty feet below the bed of the stream or forty feet below the zero level. The framing of the ice breaker and the bracing within the dam was of enormous strength. The number of piles driven in the four coffer-dams reached the enormous total of 5,224, and of the 1,227 driven in dam No. 3, $16\frac{1}{4}$ per cent. were drawn and redriven. These piles and the timber were obtained from the forests of Bavaria and Upper Austria. Fig. 7.

The first pile on dam No. 3 was set on April 8, 1842, but owing to the difficulties encountered it was not finished until three years later—April 4, 1845. From six to seven days were occupied at the first in driving a pile to a depth of five or six feet into the clay, but as the work progressed the difficulty increased, the operation of driving one pile consuming from twelve to fourteen days, many piles breaking short off so they could not be withdrawn, and the gravel was dredged out from behind and a second row driven. The report further describes the difficulty of the work: "The dredging for the No. 3 dam was carried on to the average depth of forty-four feet from the top of the outer row of piles, leaving about ten feet of gravel to drive through, and extra piles were driven where the gravel found its way between the piles, as well as where it was known the piles were not driven to the proper depth, or were broken or otherwise injured. As the gravel was dredged out to the above depth, the inner and middle row of piles were driven, and a great part of them got down as was supposed to the requisite depth. The work was carried on in the above manner until the 7th of November, when from the appearance of several piles which were pulled up, and from other causes, it became apparent that the outer row was in a much worse state than had been expected and was almost a matter of certainty, that those piles which had taken ten or twelve days to get down were not driven to the proper depth by at least three or four feet, having upset or lost their points to that extent. There was likewise every reason to believe that many of them were broken or dangerously crippled. Added to this the Danube was rising, and at the late time of the year, with winter rapidly approaching, the general appearance of the dam was anything but satisfactory. Upon mature consideration the only course appeared to be to drive a much greater number of piles than was at first calculated upon, and another complete row of piles was driven all round at intervals of fifteen inches apart, and in some cases double and triple piles were driven during the progress of

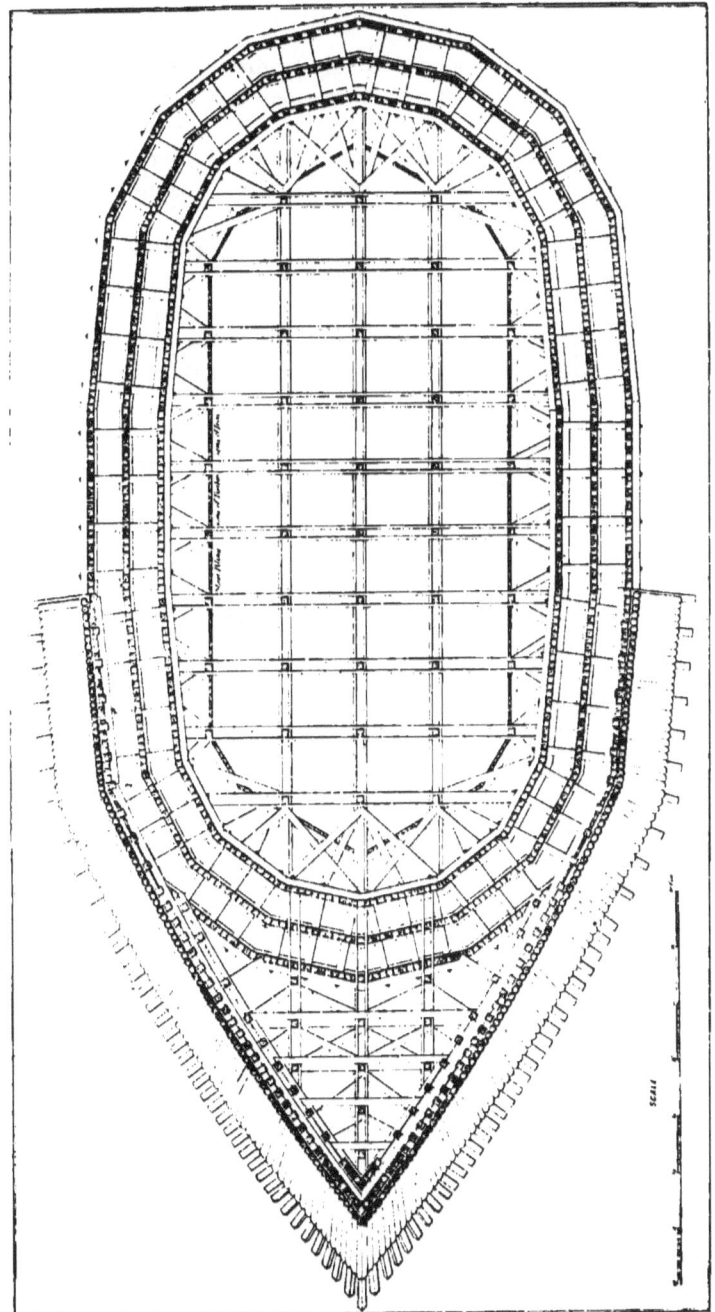

FIG. 7.—BUDA PESTH SUSPENSION BRIDGE. PLAN OF COFFER DAM NO. 2.

the dredging. At the commencement of the driving a few were got down to the depth of fifty-seven or fifty-eight feet, being from three to four feet in the clay; but as the gravel began to get compressed many of them would not penetrate more than fifty-four or fifty-five feet, the sharp angular gravel overlying the clay appearing to be compressed into a substance as hard as rock."

The puddle used was clay mixed with about one-third clean gravel, it having been found to set quite solid, from experiments made by sinking specimens in the Danube. When leaks occurred they were closed by driving square timbers down thirty or forty feet into the puddle to pack it or by driving new piles to close the cracks and in some cases by driving sheet piling.

Experiences of this nature led to the disuse of coffer-dams for foundations to such depths, but a very small percentage of the care exercised and the persistence shown in this work would lead to greater success on ordinary foundations.

The class of work to which coffer-dams may still be applied will be shown in the succeeding pages and the examples from actual practice will show in some measure the care that must be exercised in the first construction to prevent failure, and the expedients adopted to overcome unavoidable accidents.

"In every man's mind, some images, words and facts remain, without effort on his part to imprint them, which others forget, and afterwards these illustrate to him important laws."

ARTICLE II.

THE COFFER-DAM PROCESS FOR PIERS.

CONSTRUCTION AND PRACTICE.

THE exact definition of the term coffer-dam—"a water-tight inclosure, from which the water is pumped to expose the bottom and permit the laying of foundations"—is the class of structure which is to be considered, although in the construction of them cribs or caissons may be employed and utilized; the essential purpose being to form an inclosure as nearly watertight as possible in order that the expenditure of power for pumping out the water may be of small amount.

The attainment of this when the water is shallow and has little current we have seen to be easily accomplished by means of a bank of clay or clayey gravel.

This form may also be employed in still water up to about four feet in depth by the addition of sheet piling or a casing supported by ordinary piles to prevent the embankment from caving into the excavation. Where the bottom is of soft mud or porous material over a solid clay or gravel, as much as possible of the porous material should be removed before forming the embankment, thus preventing leakage underneath. In very shallow water this can be accomplished by shoveling and with large hoes or scoops, but with several feet of water to contend with, some form of dredge or scraper must be employed. A very convenient form of scraper used by M. L. Byers on the Cinti. & Mus. Valley Railway is described in Vol. 31 of the "Transactions of the American Society of Civil Engineers," and consists of old boiler iron, strengthened by three ribs of light iron rail as shown in Fig. 8. This was operated by a double drum 20 horse-power Mundy hoisting engine, with the towing line running directly from one drum to the scraper and the back line from the other drum over a sheave to the front of the scraper. The excavating averaged about forty-five yards of material each day during twelve days' work. The weight of the device was about one thousand pounds.

Where the material is very soft, a hand dredge called a spoon will accomplish the work at about the same cost as excavating on dry land. The spoon usually consists of a long pole, having a cutting ring fastened at one end, and to this ring is attached a canvas bag to contain the excavated material. The ring is hung from a derrick with a set of falls, being guided with the pole, as it is dragged forward by the derrick through the material to be excavated.

Excavating may be done on all the larger rivers by employing the sand or gravel diggers which are most always to be found, the dredging being accomplished by means of a series of buckets on a belt or on chains operated through a well in the bottom of a barge. Dredging by machinery on a large scale will be considered later on in some detail.

The method of embankment is sometimes employed for greater depths than four feet and in some instances successfully.

The Chanoine dams on the Great Kanawah River required substantial foundations beneath the water, and to accomplish this Addison M. Scott, the resident engineer, employed log cribs about the spaces, with earth banked up on the outside. This work is described in the report of the Chief of Engineers for 1896.

The site of the navigation pass of dam No. 11 including the center pier, required a coffer-dam 90 feet wide and 330 feet long inside. (Fig. 9.) This area including the necessary room for the cribs, was dredged out to hardpan from 20 to 24 feet below low water. The log cribs which contained about 84,000 lineal feet of logs, were sunk in sections 19 feet wide and 20 feet long. They were sheathed up to about three feet above low water, with sheet piling in three layers, on the Wakefield system. The driving was accomplished by attaching an eighty-pound weight to an Ingersoll-Sergeant drill run by steam and utilizing the reciprocating motion by attaching the drill with clamps to the tops of the sheathing, following it down as it was driven, after the manner of the Nasmyth steam pile hammer. This device, which is one of

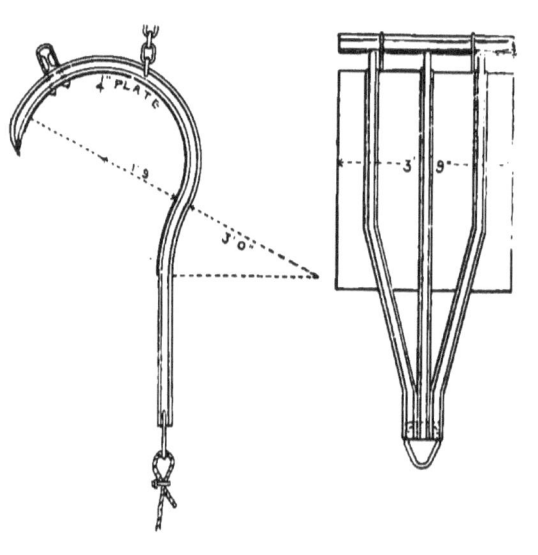

FIG. 8.—SCRAPER DREDGE.

the most ingenious ever devised for the purpose, was arranged by the contractor's engineer, S. H. Reynolds, and was a complete success.

The tops of the cribs were ten feet above low water, and the bottoms rested on the hardpan, making a total height of from thirty to thirty-four feet.

The cribs were filled with sand and gravel that had been dredged out,

FIG. 9.—COFFER-DAM AT DAM NO. 11. GREAT KANAWAH RIVER.

but the outside was banked up with selected clay and dredged material, which was protected by a layer of riprap up to about low water.

When the coffer-dam was first pumped out several leaks were developed, but after one week in perfecting the details the pumps were started regularly and no serious trouble was had afterward. This is only one of a series of coffer-dams which have been constructed on the several dams in this river, and owing to the care exercised good results were obtained uniformly.

The construction of a similar piece of work on the Ohio river was begun by Major R. L. Hoxie, corps of engineers, and is described in the report of 1895: "It was originally planned to enclose the site of the dam and lock within a coffer-dam, and work was commenced upon that basis. But on attempting to pump out the inclosure, it was found that water came in in large quantities, not only under the dam but from springs in the bottom, and all attempts to close these by dumping clay and gravel was a failure. The area inclosed by the dam was about 600 by 200 feet or about three acres of river bottom. The deposit of sand and gravel overlying the rock was about thirty-five feet thick, the rock being forty-five feet below the water level, while the plans required an excavation twenty feet deep below this water surface. The bottom deposit had been worked over for years by sand-diggers who threw back the large stones and coarse gravel after removing

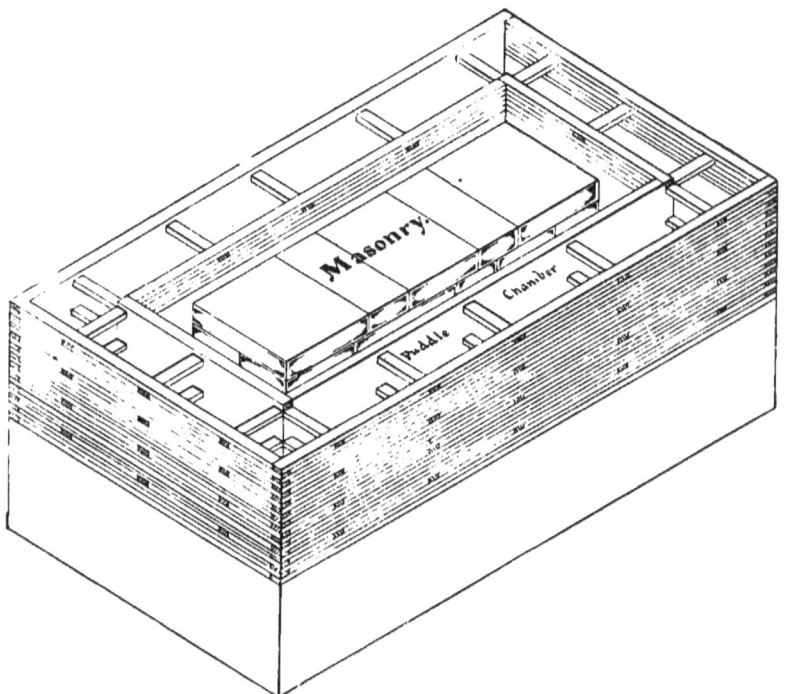

FIG. 10.—CRIB COFFER-DAM; CHICAGO, BURLINGTON AND QUINCY RAILROAD.

the fine sand, this work resulting in a very permeable bottom, with possible channels of comparatively large dimensions extending to unknown distances beyond the limits of the coffer-dam."

This is perhaps the most frequent source of failure of a well constructed coffer-dam and should always be guarded against by removing as much of the porous material as possible, by dredging before the construction of the coffer-dam is begun.

Cribs are very easy to construct, usually very substantial, and easy to make use of by floating to position and then sinking in place. A very simple form that has been used on the Chicago, Burlington & Quincy railroad is described by E. J. Blake, chief engineer. Where the water is shallow they have been built in the form shown (Fig. 10) of fence boards spiked one piece on another; with deeper water they are made of heavier timber 2"x8" or 2"x10". They are built on the water and are tied across at intervals by pieces spiked through the wall, which pieces should be carefully fitted to prevent leakage. In some cases where the bottom is soft, instead of dredging, a bottom is added to the crib to prevent the filling from squeezing its way out from under the edge.

When the crib has reached bottom, being sunk by weighting it down if

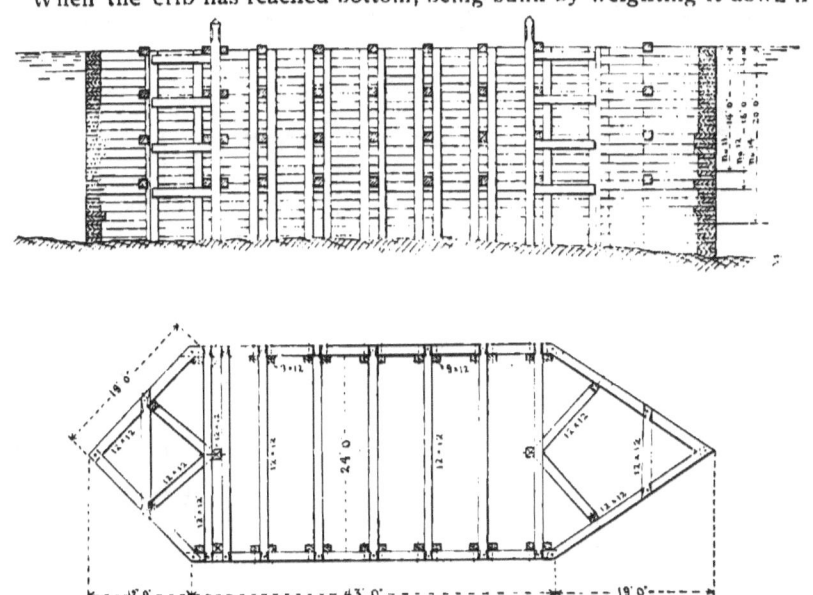

FIG. 11.—ST. LAWRENCE RIVER BRIDGE CRIB AND COFFER-DAM, CANADIAN PACIFIC RAILWAY.

necessary, the chambers are filled with clay puddle and clay is banked up around the outside to prevent water running under. The crib is made large enough so that the excavation will leave an easy slope to the inner edge of the timber work. This form can be made to conform readily to the contour of the bottom by starting the layers of timber at different elevations. No leakage has been experienced except what can readily be kept under control with ordinary sized centrifugal pumps. The cost of construction is generally a minimum, as there are usually plenty of old timbers available for use from the railroad yards.

Cribs constructed in a similar manner but with only one wall of timber have been used successfully on the Canadian Pacific Railway by P. Alex. Peterson, chief engineer.

The bracing is very efficiently attached by dovetailing it into the sides, while the form of the crib enables it to withstand the force of the current and the ice. The projections on the inside are to prevent the water from forcing its way up between the sides and the concrete filling when the dam is pumped out. These projections answered their purpose very effectually, and when the dam was pumped out it remained dry enough to lay the masonry without any additional pumping.

Illustrations are given of a crib of this character which was used on the St. Lawrence river (Fig. 11) similar ones being used for the other piers of the same bridge, and of the crib used for the Arnprior bridge. (Fig. 12.) This shows the concrete which was deposited on which to found the masonry, and

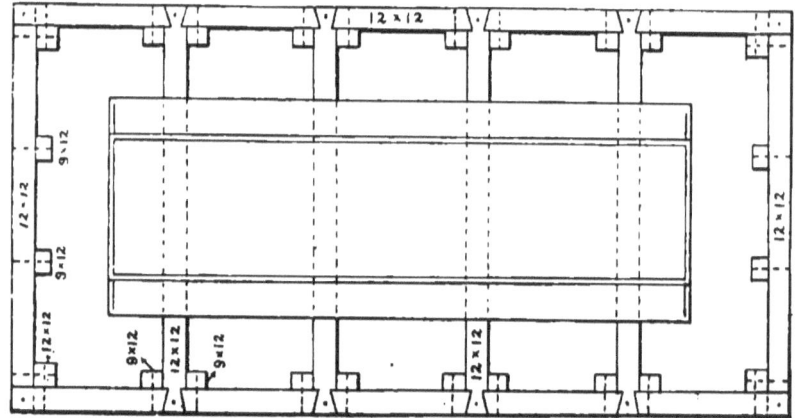

FIG. 12.—ARNPRIOR BRIDGE CRIB AND COFFER-DAM, CANADIAN PACIFIC RAILWAY.

which formed a watertight bottom so that the crib could be pumped out for the laying of the stone.

The practice on the Atchison, Topeka & Santa Fé railroad has been in some respects similar to what has been given. C. D. Purdon, assistant chief engineer, states that cribs built of old timbers are used when such material as stringers 7"x16" is plentiful, each course being stepped in about one-half an inch to give a batter. For use in sand when rocks and drift are likely to be encountered a crib is made by constructing a frame of old bridge timbers and sheathing it with plank. (Fig. 13.) This is sunk by digging out the sand, which is shoveled first into box A, then to boxes B, then to C, and then outside. The suction pipe is shown in dotted lines, the pumping being accomplished with a centrifugal pump. This plan works very suc-

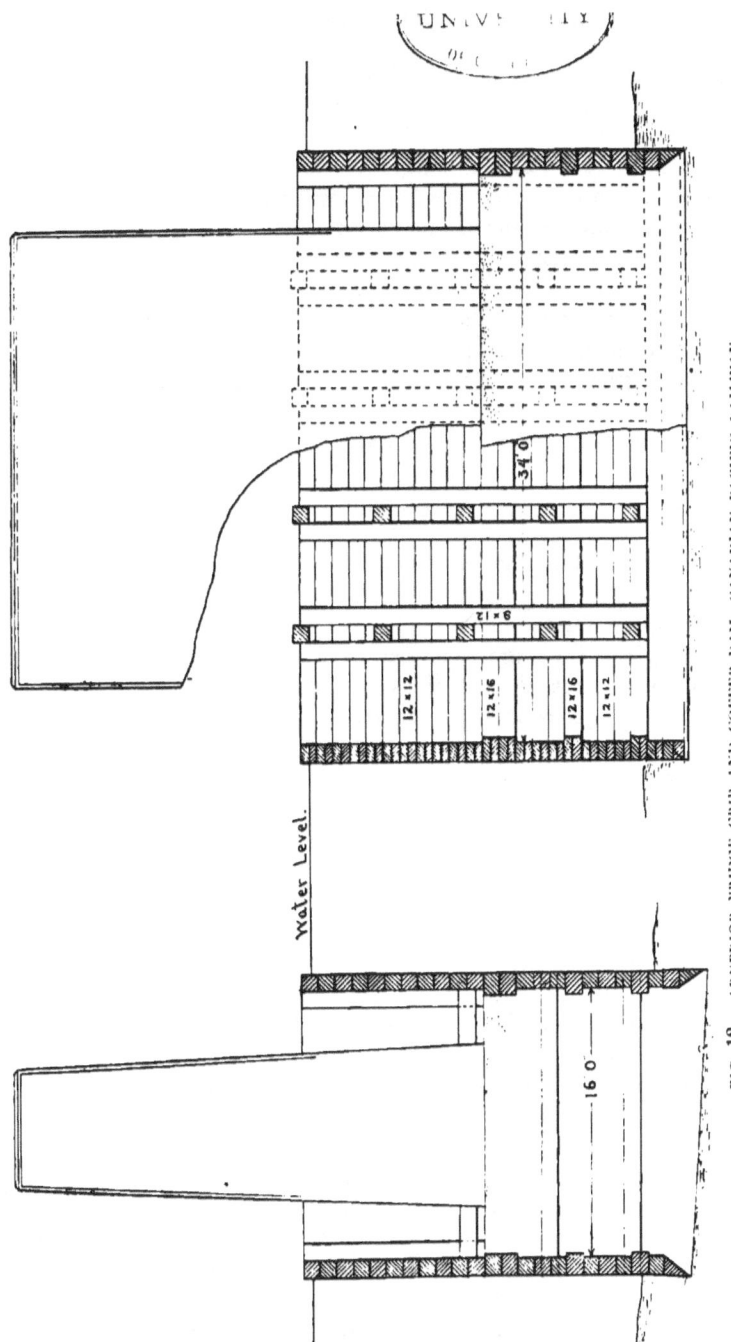

FIG. 12.—ARNPRIOR BRIDGE CRIB AND COFFER DAM, CANADIAN PACIFIC RAILWAY.

cessfully on the streams in Colorado and New Mexico where the water is mostly in the sand and but little shows as surface water.

The Arkansas river bridge of the St. Louis & San Francisco railroad at Tulsa was built over a bottom of gravel and riprap above rock, which was quite level and about seven feet below low water. Cribs were constructed for coffer-dams similar to the one just described and set on the bed of the stream. Clay from the bank was dumped outside and as the crib was dug out and sunk, the clay followed down and kept out the water.

When the bottom is of clay or of sand without obstructions, sheet piles, either tongue and groove or the Wakefield, are driven around a crib.

Geo. H. Pegram, chief engineer of the Union Pacific system, has made the construction of coffer-dams conform to available material and local conditions. At the crossing of the Republican river in Kansas, where the bottom was sandy, a single thickness of four-inch V-shaped tongue and groove sheet-piling, with the usual guide piles and wales, served to form a watertight structure.

Where a gravel bottom overlaid a hard soapstone, as on some work in Idaho, with seven feet of water to contend with, the coffer-dam was made of Wakefield piling, formed of 1½-inch sized plank. The joints were tightened with cement; and sand, gravel and straw placed outside to prevent leaking. Wakefield piling has also been used for clean rock bottom, placed in two rows about the depth of the water apart. Intermediate cribs filled with rock were used to sink them. The ends of the piling were sharpened and driven on the rock until broomed up and rendered nearly watertight, when gravel mixed with straw was placed around outside to close any remaining leaks.

In cases where ordinary piling has been driven and a grillage laid upon them to receive the masonry, a coffer-dam is constructed as shown (Fig. 14) in which to lay the masonry. The construction of this is fully shown in the different views given.

Another form of coffer-dam for the same purpose was constructed by Octave Chanute in laying the masonry of the pivot pier for the Fort Madison bridge over the Mississippi river, on the line of the Atchison, Topeka & Santa Fé railroad. (Fig. 15). This is described in the *Engineering News* of June 2, 1888, by W. W. Curtis, resident engineer : "The grillage (for the pivot pier) is four feet, three inches thick, the upper fifteen inches being dressed to an accurate circle of the desired diameter. The coffer-dam was footed against these two courses and was formed of 3"x8" pine plank staves, dressed on the sides to a slight bevel around which were placed seven wrought iron hoops $4"x\frac{3}{16}"$, $5"x$ ", and $6"x\frac{3}{16}"$, similar to those used for water tanks, and screwed up tight. Inside of these, circular braces of plank were fitted. As a water pressure of nineteen feet was to be resisted, additional security against leakage was obtained by placing a string of candle wicking vertically between

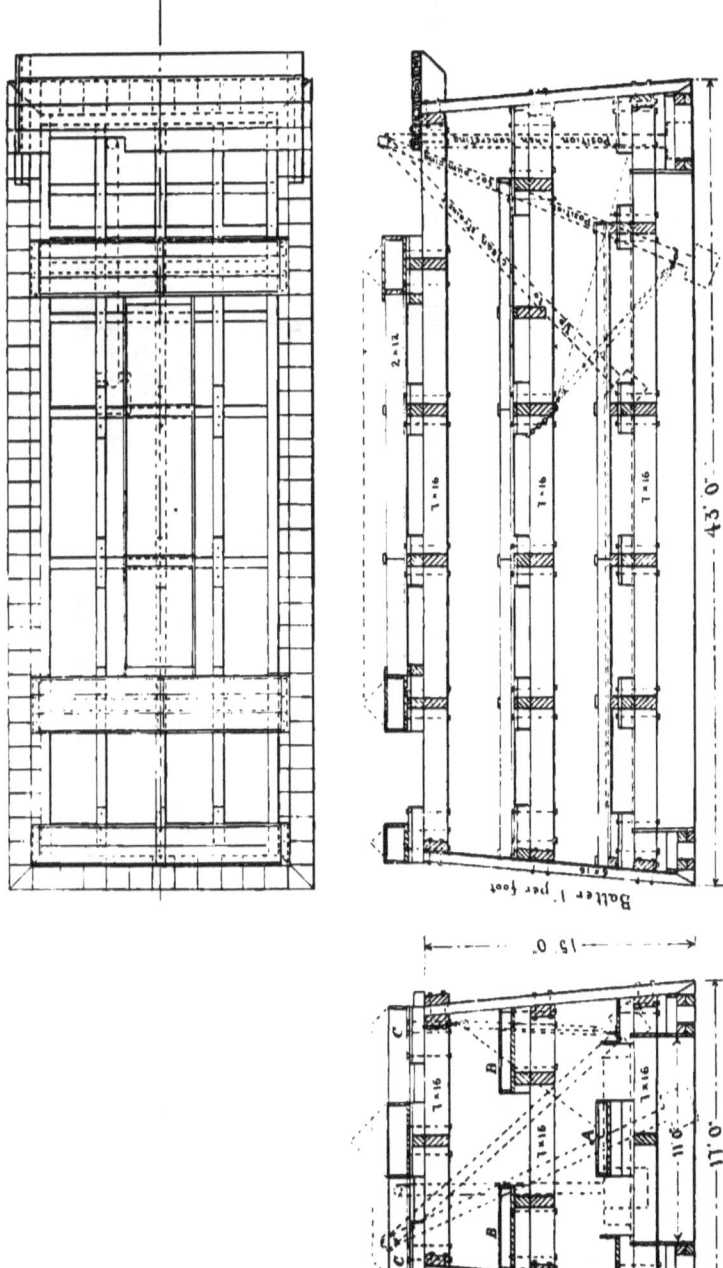

FIG. 13.—CRIB COFFER-DAM, ATCHISON, TOPEKA AND SANTA FÉ RAILWAY.

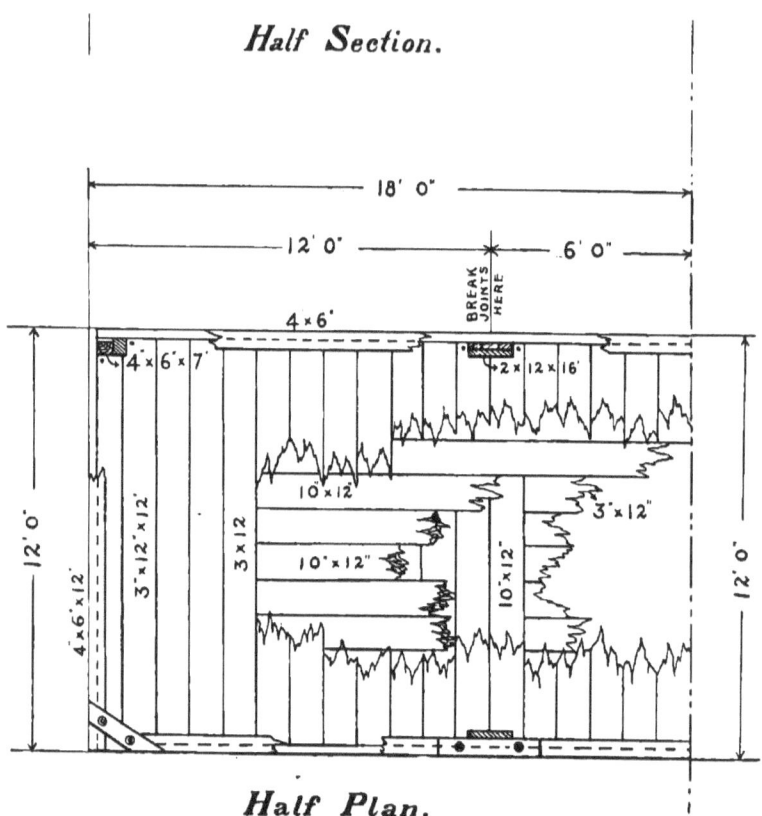

FIG. 14.—COFFER DAM ON GRILLAGE; PAYETTE AND WEISER RIVER BRIDGES. UNION PACIFIC SYSTEM

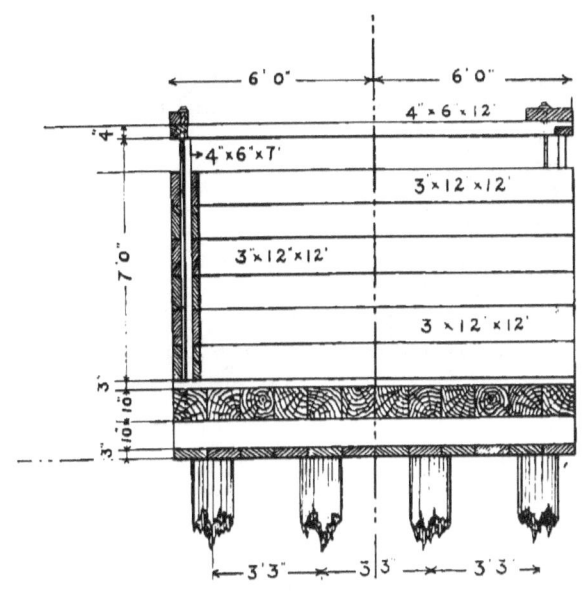
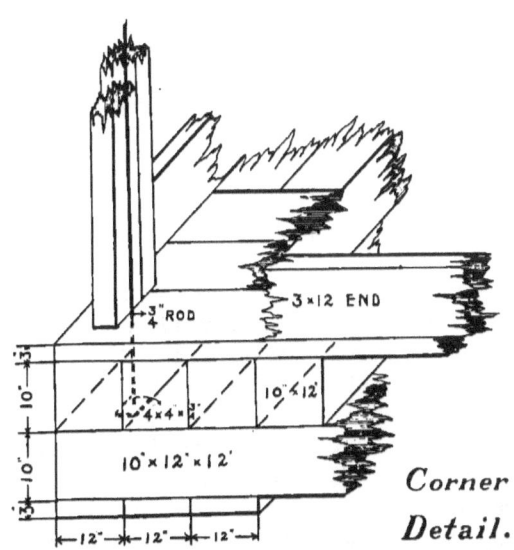

FIG. 14 —COFFER-DAM ON GRILLAGE; PAYETTE AND WEISER RIVER BRIDGES UNION PACIFIC SYSTEM.

each stave. When the caisson was submerged to about full depth it became necessary for the steamboat to assist it into final position. A 12″x12″ post was bedded in the concrete in the center of the pier, with four braces running to the circular bracing of the sides. This makes a very cheap cofferdam and was found to work very well."

An attempt to use a form similar to this was made in constructing the Walnut Street bridge at Philadelphia. This is described by Geo. S. Webster, chief engineer Bureau of Surveys, in the *Engineering News* of March 15, 1894: "In founding the river piers, the Robinson coffer-dam was first tried, but was abandoned after three of them had failed by collapsing. This

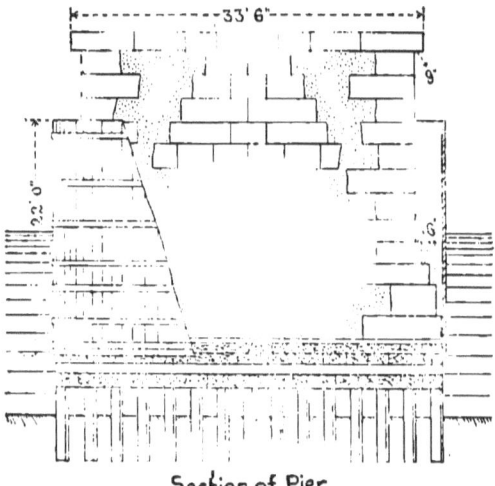

FIG. 15.—COFFER-DAM ON GRILLAGE, FORT MADISON BRIDGE, ATCHISON, TOPEKA AND SANTA FE RAILWAY.

dam may be briefly described as follows : A circular platform about eighty feet in diameter supported upon piles at an elevation of about four feet above high water was first constructed. Square piles of 12″x12″ yellow pine were then prepared by spiking a 3″x4″ timber flat, along the middle of one side, and two others along the edges of the opposite side, forming a tongue and groove on each pile. The tops were squared off and the bottom ends pointed to a wedge shape. These piles were then driven close together against the edge of the circular platform and down to rock. Mr. Robinson's idea was that the mud overlying the rock would hold the piles in position at the bottom, and if the top ends were held by an outside hoop, the dam would be secure without internal bracing to resist collapsing pressure. In the first trial the hoop was made of boiler iron some four feet or more in width. In the second dam it was formed of a heavy steel railway rail, and in the third dam the hoop was the same as in the second, but it also had a number of radial rods in addition. The first dam was pumped out and held for nearly an hour before collapsing, but the others collapsed before being entirely pumped out. After the third failure this form of dam was abandoned."

It would seem likely from a comparison of the two cases, one being entirely successful and the other a failure, that had the Walnut street dam been supplied with additional bands lower down and provided with some means of tightening, with several internal bracing ribs of timber, it would have proven a success. These bands and ribs could likely have been placed by a diver.

The uncertainty which always exists regarding any construction under

FIG. 16.- A CRIB COFFER-DAM AFTER A FLOOD.

water makes it imperative that every precaution should be taken to guard against troubles that might arise, by making the construction of no doubtful form and in no doubtful manner from its first inception.

The nature of the bottom will always indicate the method of construction which should be adopted in a given case, but it would be rarely that the preliminary dredging could be dispensed with. It is true that there are cases where there is a deposit overlying a seamy rock, and the water will find its way along the seams, bubbling up in springs inside. Resource must be had to cutting off the flow, by puddling on the outside, sometimes extending the operations a distance of a hundred feet or more away, until enough of the flow has been stopped so that the water can be kept down by a reasonable amount of pumping.

The next precaution after dredging, is the building of some form of cofferdam which shall effectually exclude any flow through the sides of the dam. This we have seen to be accomplished in many cases by means of a bank of clay, or a row of sheet-piling, and in some cases by a single walled crib. But in the last two methods a supplementary bank of clay or clayey gravel on the outside is necessary to prevent leakage.

This bank may be protected from wash by covering it with clay, sand or gravel in gunny sacks, or by riprapping up to about low water, as was done on the Kanawah dams.

Double walled cribs and coffer-dams constructed with two rows of watertight sheet piling, require to be puddled with a carefully selected material. While clay can be used with a good degree of success, it will be found better to use a clayey gravel or to mix the clay and gravel, as was done at the Buda-Pesth bridge. When a small leak starts through a pure clay puddle, it washes out the clay in considerable quantities and a dangerous leak is soon developed. With the admixture of gravel, however, a leak is stopped almost as quick as started by the heavier gravel falling into and closing the void.

It will generally be found advantageous to use a bank of clay outside of a double walled dam, unless it might be a case where sheet piling has been driven to rock, and even then a certain amount of material in sacks should be used to prevent wash or the cutting out of the earth around the sheeting.

Whatever excavation is taken out of the interior of the coffer-dam after it has been pumped, should be dumped at the upstream end and corners, or to fill any holes or pockets there may be around the sides or ends.

Cutwaters should be added to all coffer-dams which are built in rivers having a swift current or a heavy flow of ice, as was the case at Buda-Pesth and on the Canadian Pacific examples. They must also be used in rivers where the run of drift with each rise is of large amount. For the purpose of preventing wash around a dam, a cutwater of plank supported by a frame of timber may be constructed separate from the main structure, or a V-shaped row of sheet piling driven up stream. On rock, a timber crib of triangular shape, built of round logs, may be sunk up stream and filled with broken stone. Such a crib can be utilized in anchoring the main crib of a cofferdam, as was done at St. Louis, and which will be described in future pages.

More fitting language cannot be found for closing words than those used in Wellington's monumental work on railway location: "The uncertainty as to the exact requirements to be fulfilled by the works when completed is a disadvantage, indeed, which cannot be escaped; but the more difficult it is to reach absolute correctness, the greater need we have of some guide which shall reduce the unavoidable guess-work to its lowest terms, and so save us from the manifold hazards which result from not only guessing at facts, but at the effect of those facts. Whatever care we use we can never attempt with success to fix the exact point where economy ends and extravagance begins; but what we can do is to establish certain narrow limits in either direction, somewhere within which lies the truth, and anywhere outside of which lies a certainty of error."

ARTICLE III.

THE COFFER-DAM PROCESS FOR PIERS.

CONSTRUCTION AND PRACTICE.

WHEN for some reason the necessary care has not been exercised in the construction of a coffer-dam and in puddling it, or where there were discovered conditions not known before the construction began, which rendered the work unsatisfactory or leaky, it will usually be found that the mode of repair which seems most expensive will in the end prove the cheapest and most expeditious. If the puddle proves leaky, and it be decided that the material was of too porous a nature, the best remedy is to dig out and replace it with better. Should it be found that the porous bottom had not been removed to a sufficient depth, it may be found necessary to dig out the puddle chambers and puddle deeper, or the leaks might be stopped by banking up outside of the dam with clay or clayey gravel, or perhaps sand in sacks would do some good.

Gravel will allow the percolation of water even where the head is small, and when a pressure of from four feet upwards is brought upon it, the leakage becomes considerable and difficult to control, so that pure gravel is of little service in stopping leaks.

Hay, straw, oats, crushed cane stalks, rotten stable manure, and similar materials, mixed with the banking material, are very efficacious in producing tightness, and when applied to local leaks will assist in closing them.

Where sheet piling have been used to exclude the water and leaks still occur, they can often be closed by driving more sheeting to lap the cracks, which may have been widened out lower down as the sheet piles were first driven. This, we have seen, produced satisfactory results at Buda-Pesth, where leaks were also closed by driving square timbers into the puddle to compact it.

Clay can also be forced down through pipes directly to where the leakage occurs. The use of this at the Government Lock at Sault Ste. Marie is described in the *Engineering News* of September 26, 1896: "The only difficulty encountered in the work of excavation was due to a leak in the cofferdam, which flooded the lock pit and delayed the work considerably. The cause of this leak was found to be a crevice in the rock passing underneath the coffer-dam, and despite all efforts to close it, the flow of water rapidly enlarged the break until about fifty feet of the clay in the coffer-dam had been washed away. The large break was closed by driving additional sheet piling and filling in with brush, hay, and clay in sacks. This, however, failed to entirely stop the leak through the crevice, and it was determined to fill the cavity with clay. For this purpose a 3-inch pipe was driven down through the coffer-dam until its lower end penetrated the crevice. In this

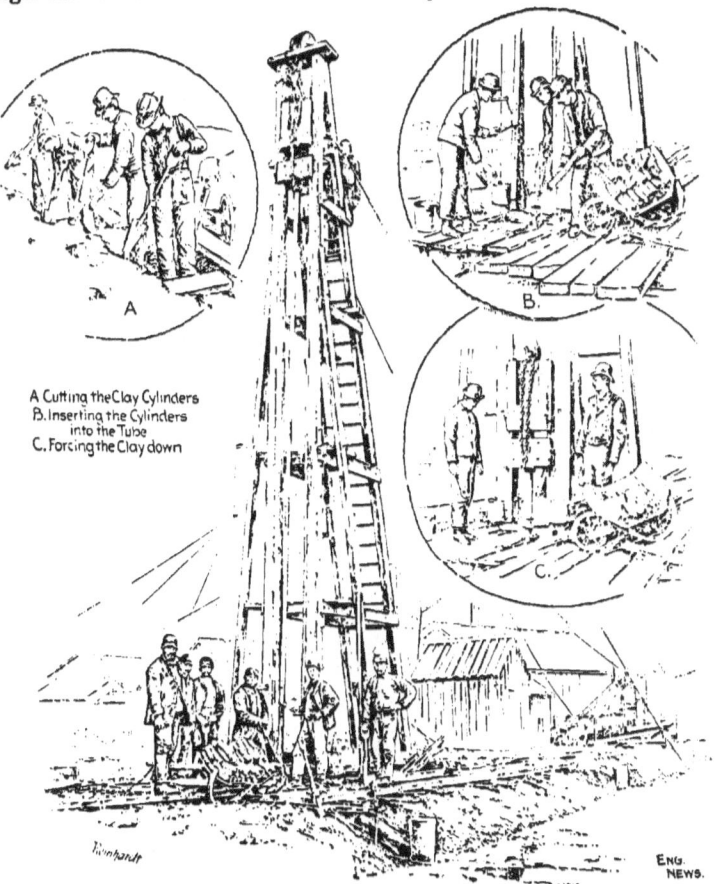

A. Cutting the Clay Cylinders
B. Inserting the Cylinders into the Tube
C. Forcing the Clay down

G.17. APPARATUS USED TO FORCE CLAY INTO CREVICE OF FOUNDATION ROCK AND CLOSE LEAK IN COFFERDAM.

pipe small cylinders of clay about one foot long were placed and forced down into the cavity by means of a plunger working in the pipe. The apparatus is shown in the illustration (Fig. 17). As will be seen, the plunger, or rammer, is an iron rod to the top of which is fastened a block of wood sliding between the guides of an ordinary pile driver. The hammer of the pile driver is the weight which pushes down the rammer. This apparatus was designed by E. S. Wheeler, engineer in charge of the work, and was used not only to fill the crevice, but all along the coffer-dam for the purpose of compacting the clay filling. The apparatus proved most successful for the purpose for which it was intended."

The use of rods for bracing in double walled coffer-dams is very often the cause of considerable leakage, the water following along them through the puddle. This may be stopped by wrapping a band of hay or straw around the rod next to the timbers, or by a wrapping of coarse cloth, or by a wood washer having a hole slightly smaller than the rod, which is forced through.

The walls of the dam must always be made tight, and this we have seen to be effected by careful framing of sides and bracing, and it will be seen in a later example how round struts between the two walls allowed the puddle to flow around them and close up much better than if the braces were square timbers.

The use of candle-wicking between the staves proved successful at Fort Madison, and calking is very often resorted to at the first, and also to close up local leaks. The use of this and the use of a stiff grease between the layers of a crib will be referred to in another part of this article.

The use of tarpaulins to make a watertight piece of work is described in the Trans. Am. Soc. C. E., Vol. 31, by Montgomery Meigs, engineer in charge of the government work at Keokuk, Iowa. "The upper one of three locks was twice repaired by separating it from the river by an ordinary plank and mud coffer-dam. But as this work had to be done after the close of navigation, it was found to be very unsatisfactory on account of the freezing of the puddle, and on one occasion the partly puddled dam froze and upset. After this experience it was determined to use some other method than puddle to produce tightness. There was available for drainage a 50-H. P. suction dredge, with 14-inch suction, and a rotary Van Wie pump, and plenty of 12-inch discharge pipe mounted on pontoons. It was proposed to drain the lock with this dredge, allowing the boat to settle in the mud at the bottom of the lock as the water left it, and to complete the work with a 3-inch discharge Pulsometer. The lock being 350 feet long and 80 feet wide, a flat place on the bottom was selected, the dredge placed over it and the necessary length of discharge pipe placed in position on its pontoons. The point selected for a bulkhead (Figs. 18 and 19) was just outside the lock gates, about forty feet below the lower mitre sill, where there was a smooth rock bottom, the ends of the dam abutting against the flaring ashlar wing walls of the lock approach.

"The bulkhead was constructed with thirteen bents eight feet apart, of the size timber shown, with light diagonal bracing. After being built 2½ miles from the lock it was towed to position and sunk by weighting it with old railroad rails, enough being used to overcome the buoyancy after the sheathing was added. A diver was employed to see that the bottom was clear of obstructions and to guide the bulkhead to a solid bearing. The sheathing was also guided to place by his assistance.

"The canvas sheet, which was designed to give tightness to the apron, was of two breadths of ten feet and one breadth of six feet wide, sewed together edge to edge for convenience, and about four feet longer than the

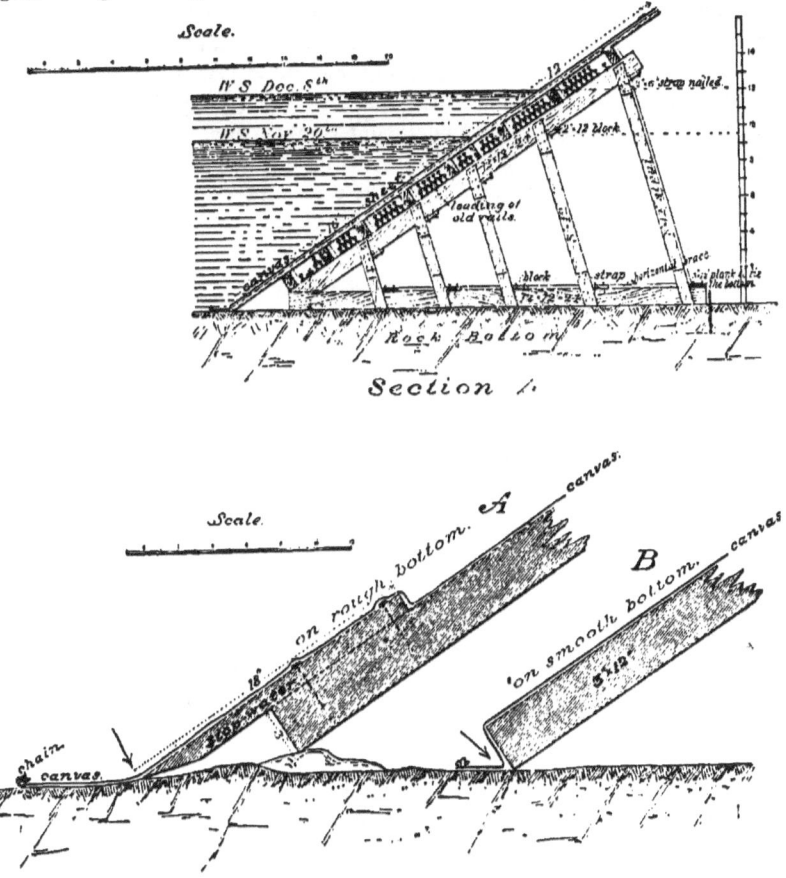

FIG. 18.—DETAILS OF CANVAS AND PLANK BULKHEAD.

extreme length of the apron. Some old ½-inch and ⅝-inch chain was sewed to one edge continuously to act as a sinker and insure the lower edge of the canvas sheet hugging the bottom tightly. A few stones laid on it would have answered the same purpose, but not so well. The canvas was 12-ounce duck.

"The sheet was spread under water by the diver. It lapped on the bottom about twelve inches, covered the face of the apron and extended some inches up the face of the wing walls at the end of the dam. Cleats were nailed on the angle between the apron and the wing walls. These were of 1x4-inch strips, nailed with 2-inch wire nails about twelve inches apart. The upper edge of the canvas was also lightly cleated to the planking in a similar manner. No other nails were driven in the canvas, which was designed to be cut up into tarpaulins eventually. Where the plank touched bottom no beveling was used, but one ragged hole was stopped with the beveled "stop waters" which were made use of. The dam was pumped out in about six hours and the leakage was so small that a 3-inch discharge pulsometer kept out the water, and was then run only at intervals. Small leaks were stopped by dumping rotten stable manure in their vicinity."

It is interesting to note that the bulkhead stood a pressure of twelve feet of water. Experiments made to determine what pressure 12-ounce duck would stand, show that the clean canvas begins to leak at two pounds pressure, and at five pounds pressure the leakage becomes a marked amount. With mud on the canvass the leakage becomes noticeable at from five to seven pounds, and of a considerable amount at fifty pounds pressure, these pressures being on a circle $4\frac{1}{2}$ inches in diameter. The canvas did not rupture at 800 pounds.

The suggestion is made to use an inverted funnel of canvas to stop the leakage of springs on rock bottom. (Fig. 20.) The canvas to be spread out over the bottom and weighted down with concrete, and the top wired to a pipe into which the water may rise until the pressure head is overcome or the pipe can be plugged. Arrangements of this nature, but without the canvas funnel, have been frequently used. An iron pipe set on end is fitted over the leak, and after concreting around to make it watertight, the water rises inside until the pressure is balanced. A watertight wooden box may also be used for the same purpose.

The founding of a new inlet tower in the Mississippi at the St. Louis water works was accomplished by using a coffer-dam and it was the intention to form a junction with the bottom by using a canvas curtain. When the coffer-dam was floated into position and the divers were sent down to spread the canvas and weight it down with stones, it was found to be damaged so as to be useless. This was supposed to be due to the action of the swift current, but was most probably due to some accident such as fouling on a snag or against a barge.

The anchoring of the crib for this dam is related in the *Engineering News* of July 4, 1891. The dam was to be located near the head of a stone dike about twenty feet in height and on solid rock bottom which was uneven and worn into grooves by the action of the current, which had a velocity of between six and eight miles per hour. The bottom was leveled off by blasting, to receive the crib, which was to be sunk in from fifteen to eighteen feet of water.

The three triangular cribs shown (Fig. 21) were sunk and filled with stone and were used to hold the dam in place while building and while being sunk. Steel cables 1½ inches in diameter were used as anchors.

The large crib also served as a protection from the current and drift.

The size of the crib was 38x74 feet outside and the height 22 feet. The 12x12-inch yellow pine timbers were drift-bolted together with from one to two feet spacing of bolts, and all the joints between the timbers were calked. The bracing consisted of 12-inch square timbers, of which there were three rows, the braces in each row being four feet apart vertically. These were cut out as the masonry was built up and bracing against the stone work substituted.

There were four sets of diagonal bracing as shown. The space between the walls, which was three feet, was partly filled with concrete in sacks, and puddle placed on top. Sacks of clay were banked up around the outside, and then the dam was pumped dry with a 10-inch pump. Inside was found eight feet of mud and sixty sacks of concrete which had been washed there by the swift current.

The amount of timber used was 125,000 feet, B. M., and about 12,000 feet of ⅞-inch round iron for drift bolts. The puddle chamber required 1,000 sacks of concrete and 100 barge loads of clay, while 10,000 sacks were used for banking up clay on the outside. This work was constructed under the direction of C. V. Mersereau, Division Engineer, under S. B. Russell, Principal Assistant Engineer.

The Queen's bridge at Melbourne, Australia, is a plate girder structure, with four piers of eight cylinders each. The bottom was a reef of bluestone which had been shattered by blasting and which was silted over with about three feet of very soft silt.

The use of ordinary puddle coffer-dams was thought to be too expensive as the bridge was 100 feet in width, and it was proposed to use a single wall of timber protected by tarpaulins. The account of this work is taken from the *Engineering News* of April 4, 1895, which is an abstract of a paper by W. R. Renwick, engineer in charge.

To insure as light a construction as possible experiments were made on the strength of Oregon pine, and it was found that tests of water soaked timber showed a loss of strength of as much as 33 per cent., when compared with tests of seasoned timber. The break, too, of the water-soaked pieces was very short. This strength being the one adopted, a very low factor of safety was used. A separate dam was constructed around each tube, but with one side to open as a door to allow its removal and use for another

FIG. 19.—INSIDE VIEW OF BULKHEAD. LOCK PUMPED DRY.

place. The frame was made from 12x12 Oregon pine, with the sticks placed closer together near the bottom to resist the greater water pressure, and 12x12 pieces were run up the corners, the frames being notched in. These also served as spacers for the side timbers and as door frames. The sheeting on the outside was of 4x12 rough timber, and outside of this at the top and bottom were wale pieces, 6x12, bolted through the frames with 1-inch bolts to hold the sheeting in place.

The tarpaulin was passed completely around the dam, being tacked to the waling pieces, and so arranged as to allow the door to open.

When the dam had been placed around a tube the sheeting was driven down to rock, through puddle which had been dumped on the bottom, and

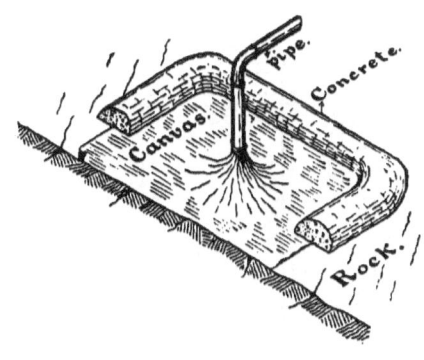

FIG. 20.—CANVAS FUNNEL FOR CLOSING LEAKS.

the pumping was readily done with pulsometer pumps. The only serious leaking was where the 1-inch bolts passed through the joints between the sheeting, but these were plugged with soft wood plugs, and in other work the bolts were flattened to three-eighths of an inch where they passed between the plank. The dams were removed by first drawing the sheeting up to its original position, when the door was opened and the crib taken to another tube. The depth of water was about fifteen feet, but while this was successful in this instance, the method should not be copied unless the conditions are favorable, nor unless the cribs are made practically watertight in themselves.

This was the case in the above work, as one of the tarpaulins was accidentally torn off and the dam still excluded the water, so that the tarpaulin was only a wise precaution. Why the cylinders were not made watertight and used as their own coffer-dam is not stated, but this possibly could have been done.

The use of tarpaulin in closing accidental leaks could doubtless be made use of frequently, but as the sole dependence for producing tightness it should be used with extreme care, in a gentle current and well protected from damage.

The pivot pier of the Harlem Ship Canal bridge was founded in a polygonal coffer-dam, from the plans of William H. Burr, consulting engineer. The work is described in the *Engineering Record* of July 24, 1897: "The rock bottom secured by the canal excavation being an acceptable surface for the masonry of the pivot pier it was constructed in a polygonal

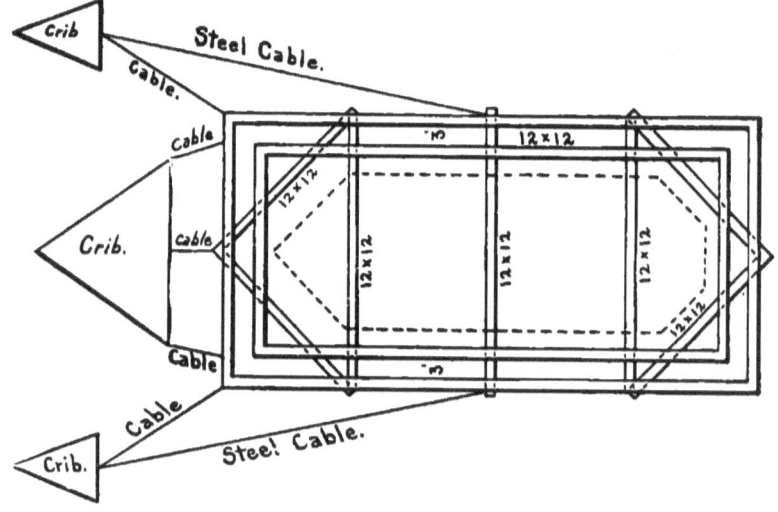

FIG. 21.—CRIBS FOR ANCHORING ST. LOUIS COFFER-DAM.

double-walled coffer-dam with thirteen sides twenty-five feet high and sixty feet in extreme diameter. The great dimensions of the coffer-dam would have made it difficult to build and launch it on shore. Consequently it was built partly on a detachable raft. As shown in the illustration (Fig. 23) the inside wall was built up of timbers lapped and halved at the angles; the outer wall timbers were carefully butt-jointed and secured by cross-struts and 1-inch bolts to the inside walls. The rough-sawed horizontal surfaces of the inner wall were bedded in stiff grease and the joints calked, which notably resisted the penetration of the water. Each course of timber was secured to the one below it by ¾-inch drift bolts spaced about four feet apart. When the bottom was thoroughly cleaned the concrete was dumped in place by a special steel bucket. Concreting was carried on night and day

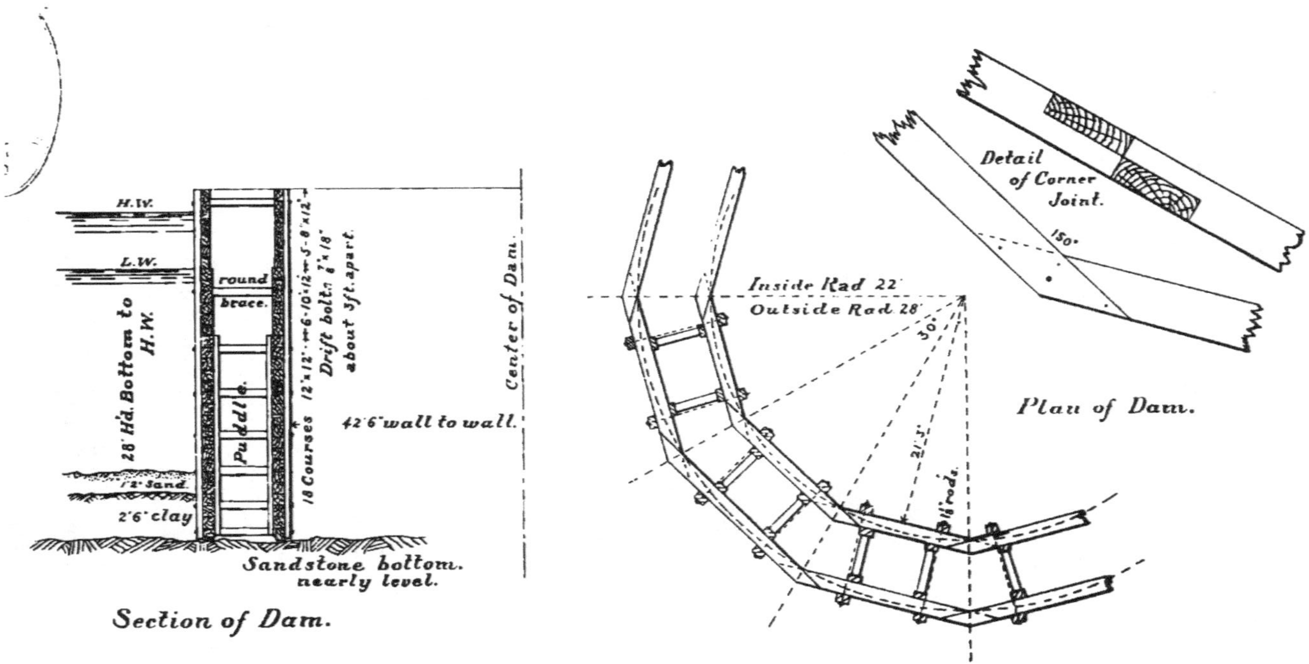

FIG. 23.—DETAILS OF COFFER-DAM USED ON ARTHUR KILL DRAW-BRIDGE.

and was completed before puddling was begun. Considerable difficulty was occasioned by the irregularities of the bottom which the coffer-dam could not be made to fit closely. Divers were sent down and filled in bags of sand, as at S, and riprap R was piled up outside to protect it. Then the space between the walls was filled with puddle."

Another polygonal dam was constructed for the draw pier of the Arthur Kill bridge, by Alfred P. Boller, consulting engineer. The following account is taken from Vol. 27 of the Transactions Am. Soc. C. E.: "It was

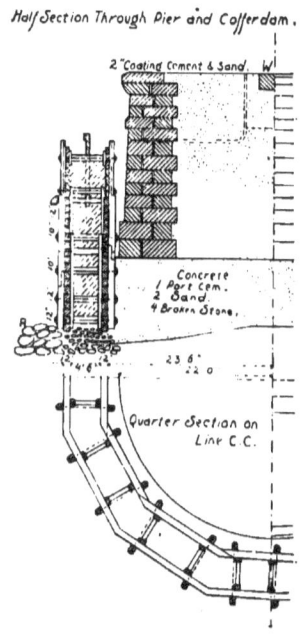

FIG. 22.—POLYGONAL COFFER-DAM, HARLEM SHIP-CANAL DRAW-BRIDGE.

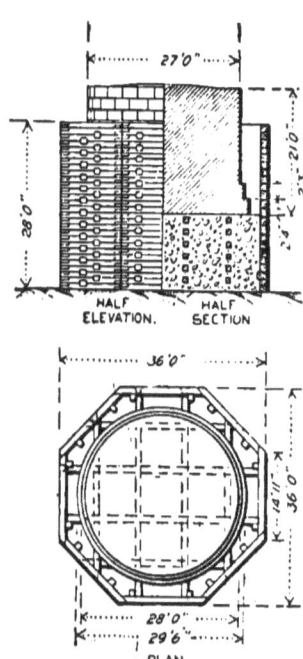

FIG. 24.—COFFER-DAM FOR PIVOT PIER OF THE COTEAU BRIDGE.

necessary to use as little space as possible for the dam, and to construct it without interior bracing, so that a double-walled twelve-sided polygon (Fig. 22) with walls four feet apart in the clear was used. The rock bottom was overlaid with two feet of clay and the clay with eighteen inches of sand and mud, the depth of water over the rock being twenty-eight feet at high tide. The square hemlock timbers used in the walls were halved together and the walls braced together by bolts and round timbers for struts, the round timbers allowing the puddle to run around them and pack well as thrown in.

Clamp timbers 4x6, in two lengths, were held in place by the bolts and the struts were braced against 6-inch plank. The dam was built to one-third its height on shore, then towed to position and built up until grounded. Between the timbers and the joints candle-wicking was placed, and the courses drift bolted together every three feet and spiked at the joints. The rock was dredged bare before placing the crib, which was filled with a hard, gravelly clay between the walls after being sunk in place. A rich Portland concrete was dumped inside, from triangular buckets, to seal the bottom and then the dam was pumped out with a 6-inch pump and kept dry by pumping at intervals. In one place the concrete was not thick enough and a spring came up through a fissure in the rock. This was boxed in and led to the sump. The material used was 140,000 feet of timber, 15,000 pounds of iron, and 600 yards of puddle."

A piece of work similar to the Canadian Pacific example was an octagonal single-walled dam used in the construction of the Coteau bridge on the Canada Atlantic Railway. This is illustrated in the *Engineering News* of May 30, 1891 (Fig. 24), and was braced thoroughly with cross-timbers built into the sides. The bottom being of rock it was partly filled with concrete to make it watertight.

The different forms of sheet piling will next be taken up, together with pile driving machinery and the methods of driving both sheet and guide piles. After this will be described the use of sheet piles for forming watertight coffer-dams, by reference to actual constructions of that character.

ARTICLE IV.

THE COFFER-DAM PROCESS FOR PIERS.*

PILE DRIVING AND SHEET PILES.

IN no department of engineering have ancient methods been more rigidly adhered to than in that of pile driving. The form of the pile-driver derrick has remained so characteristic that a person but slightly familiar with the subject would have little difficulty in recognizing the pile driver in the picture of Cæsar's Bridge (Fig. 3) in the first article. The bridge of the Emperor Trajan over the River Danube is an instance of the early use of piles. This bridge was constructed in the first century, and when the piles under water were examined in the eighteenth century they were found in some cases to have become petrified to a depth of three-fourths of an inch from the surface, beyond which the timber was in its original state. Before derricks were used it is probable that piles were driven by a large maul of hard wood, which is termed by Cresy a "three-handed beetle." The block of hard wood was hooped with iron and had two handles radiating from its center, to be worked by two men, while a third man assisted in lifting it by means of a short handle opposite.

Wooden mauls are still used where sheet piling is to be driven into a soft bottom, and heavy iron mauls or sledges are also used; but as has been frequently stated such a soft bottom should be dredged and some more elaborate apparatus used to drive the piles into a harder substratum.

The most primitive form of the pile-driving derrick is similar to the one used in 1751 by the celebrated French engineer, Perronet, at the brdige of Orleans (Fig. 25). This was arranged with a number of small ropes splayed out from the end of the lead line, so that a number of men could pull down at one time, the drop of the hammer, of course, being limited by the reach of the men's arms. The windlass shown was for the purpose of raising the pile into place between the leads.

The same engineer improved upon this derrick by adding a large bull-wheel to the windlass, on which was wound a rope to be pulled by a horse from the side, as shown in Fig. 26, thus winding up the lead line on the

---

* The subject of pile-driving has been restricted to the ordinary methods and operations; such unusual processes as gunpowder pile driving and the like have not been referred to.

Pile-driving, with the assistance of the water-jet, has been described on page 70, in the account of the Sandy Lake coffer-dam. The ordinary operations of pile-driving, as practiced on that work, are also described in some detail.

windlass. This same apparatus is in use down to the present time, except that one seen recently had the windlass at right angles to the one illustrated.

The ram or hammer used in olden times was of oak, bound with iron, and weighed for the work at Orleans 1,200 pounds for the main piles which were nine to twelve inches in diameter and which were driven three to four feet apart, center to center, to a depth of six feet into the bed of the river; the ram for the sheet piles only weighed half as much, the sheet piles being about twelve inches wide by four inches thick.

At the bridge of Saumur, which was built about the year 1756, De

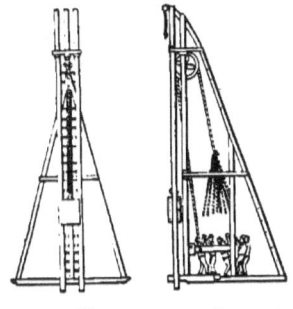

FIG. 25.—PERRONET'S PILE DRIVER.

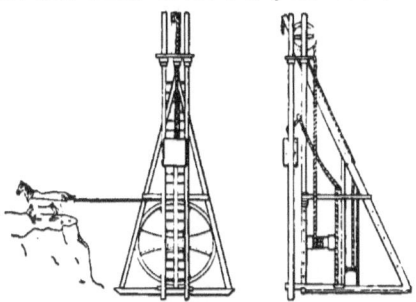

FIG. 26.—PERRONET'S BULL WHEEL PILE DRIVER.

Cessart employed a driver with a bull-wheel, in the periphery of which were set pins, to form handles for the men to pull upon and rotate the wheel. Eight men, by making three turns of the wheel, raised the ram weighing 1,500 pounds six feet, when it was unhooked and allowed to drop. The piles cost from two to five dollars each in place.

A very simple form of pile driver is shown in Fig. 27 and was described in the *Engineering News* of March 16, 1893, by Julian A. Hall. The hammer is hewed out of a section of a hardwood log, and has pieces bolted on the sides to hold it in the leads, which should give plenty of clearance. The derrick was constructed of very light timber, the verticals being 4-inch sawed stuff and the bottom timbers 6x6 inches. The rope passes over the sheave A and down over the tops of the steps B B, on which the

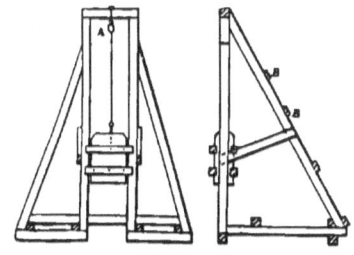

FIG. 27.—SHEET PILE DRIVER.

men stand to pull the line and thus operate the hammer. This was a very inexpensive apparatus and was found to work well. Where there is already in use a heavier hammer of cast iron it can be used by striking light blows. The construction of the ordinary pile driver derrick is a simple piece of framing, when good straight timber is easily obtained, the essential features being to keep the leads free from any obstruction for the hammer and to have efficient bracing.

For bracing a derrick under twenty-five feet a straight-back brace or

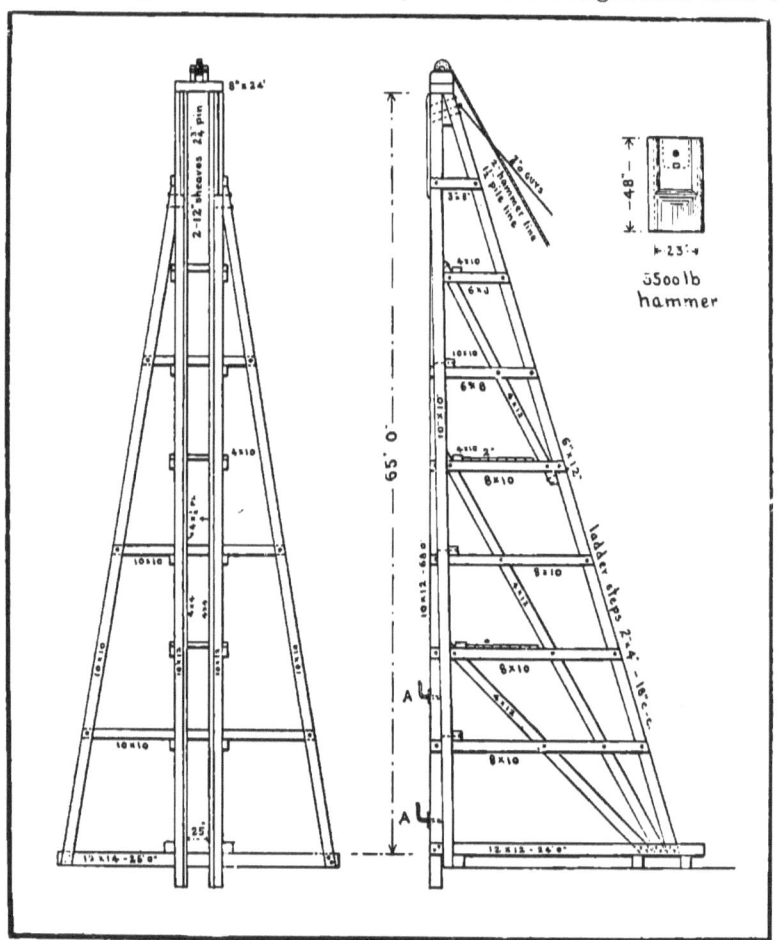

FIG. 28.—PILE DRIVER DERRICK FOR USE ON A SCOW.

ladder having two horizontals running to the leads, and two side-braces will be sufficient. But for a higher one, either additional long braces should be used or diagonals introduced between the leads and the ladder. The use of long braces is shown in Fig. 28, which is the design of pile-driver such as is used about harbors or rivers on heavy work. It would be mounted on a scow or flat-boat sixty feet in length, twenty-five feet in width and of about six feet in depth. The design of smaller derricks can be approximated from this one, the bracing being used in proportion.

It will be noticed that the guides for the hammer are 4x4 inches lined with a steel plate. Two lines are provided, one being for the operation of the hammer and the other for pulling piles into place. Especial attention is called to the hooks at A, as these are seldom shown in the plan of a derrick and they are of constant use for clamping and guiding piles. A timber laid across is wedged tight against the pile to draw it to line, and can be used to correct a stick which is beginning to slant badly. Similar clamps of course are used on the opposite side of the leads.

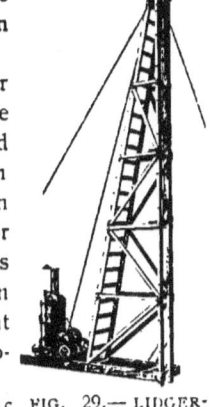

FIG. 29.— LIDGER-WOOD PILE DRIVING DERRICK.

Where a pile begins to sliver or split in driving, if the sliver is spiked down and the clamps used to hold it in place, the trouble can usually be corrected before the pile is badly damaged.

The use of diagonal bracing between the leads and ladder is shown in the Lidgerwood derrick (Fig. 29) in which a diagonal is introduced between each pair of horizontals. This form of bracing is very satisfactory and equally as good as the other method. The diagonals on a very large driver may be extended over two panels and planks spiked down to the horizontals to form a platform for the workmen. In smaller derricks the diagonal bracing is most always omitted, dependence being placed in the stiffness of the leads and the bracing from the ladder and horizontals, as was done in the derrick shown in Fig. 4.

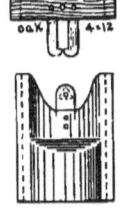

FIG. 30.—HAMMER WITH NIPPERS.

The power for driving with a small hammer weighing from 500 to 1,500 pounds, may be furnished by laborers pulling, but this is a slow operation and horse power is nearly always used where steam is not available. The power is furnished from a drum with a long lever, to which the horse is hitched and winds up the hammer by walking in a circle about the drum, the frame of which is firmly fastened in place. This is called a "horsepower" apparatus and works slowly, but is a cheap and satisfactory way where a very few piles are to be driven. To the hammer line are attached the tongs or nippers, which engage the pin in the top of the hammer (Fig. 30), and when the hammer has reached the proper height it is dropped by pulling a tripping rope and releasing the tongs, or if the hammer is hoisted to the top of the leads, the top arms of the tongs are pushed together by the wedges on the leads and the hammer released automatically. This is a slow method on account of waiting until the tongs run down again and engage the hammer. The horse power, of course, has a ratchet, so that the rope runs down free and usually the blows are hurried by overhauling the line. With the addition of a hoisting engine all this is changed and pile driving becomes one of the most stirring operations of the contractor. When the hammer is hoisted up, the friction lever is released and the hammer descends carrying the rope with it, as the tongs are done

FIG. 31.—PILE DRIVING SCOW, NEW YORK STATE CANALS.

away with and the line attached directly to the hammer. A good engine man will catch the hammer on the rebound and materially lessen the time between the blows and likewise the cost of driving.

With a heavy hammer shorter drops are made, thus causing much less damage to the pile, which would split badly under the high drop from the use of tongs. For the smaller-sized hammers—from 1,000 to 1,500 pounds—an engine of 10-horse power is mostly used, as it is usually thought best to have a surplus of power in case of need; while for a 3,000 pound hammer a 20-horse-power engine would likely prove the best and most economical, but not infrequently a 25-horse-power hoist is employed.

The cost of an outfit will vary greatly and the only satisfactory way is to get prices from responsible firms, but for preliminary estimates the cost of a 10-horse-power hoist with single cylinder and single drum may be taken at about $900, and for a 20-horse power at $1,270. Preliminary prices for other sizes of single cylinder, single drum hoists, may be obtained from the formula:

$$\text{Cost} = \sqrt[1]{81{,}000 \times \text{horse power}.}$$

The double cylinder engines will cost about 10 per cent. more and double drums about 10 per cent. additional to this.

Pile driver derricks will vary much in cost owing to the location, on account of the cost of timber, but a minimum cost for a first-class derrick will be $6 per vertical foot and a maximum of $8. Being such a simple structure the easiest and safest way will be to make an estimate for each case.

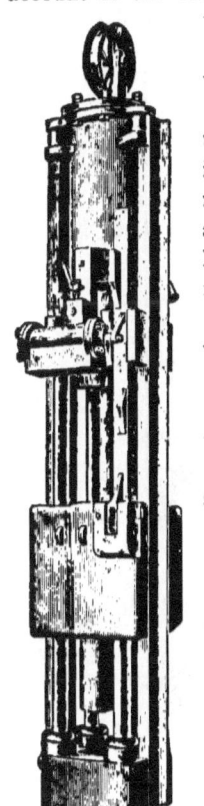

FIG. 32.—WARRINGTON-NASMYTH STEAM PILE HAMMER.

In the selection of an engine it is well to remember that with a double drum a second pile may be hoisted into place, while the first one is being driven, as all derricks are, or should be, provided with two sheave wheels at the top for this purpose. While a single-drum engine has a spool for this purpose, it cannot be used very satisfactorily.

A pile driver on a scow is shown in Fig. 31, such as was used in driving piles on the New York State canals. Another pile is just being hoisted into position. The hoisting engine has no protection, but a shed or house is mostly provided as a protection from the weather.

While little change has ever been effected in the design of pile driving derricks, the adoption of steam hoists was a great improvement, as was also the invention of the steam pile hammer by James Nasmyth. The principle is the same as that of steam forging hammers, and was applied by Nasmyth to pile driving in 1845, the hammers of this class bearing his name to-day. His idea was that the drop-hammer was calculated more for destruction than for useful effect and he termed it the "artillery or cannon ball principle." Besides this the action of the drop-hammer even with the use of the "monkey" engine was somewhat slow.

Samuel Smiles says that "in Nasmyth's new and beautiful machine he applied the elastic force of steam in raising the ram or driving-block, on which, the driving-block being disengaged, its whole weight of three tons descended on the head of the pile, and the process being repeated eighty times in a minute the pile was sent home with a rapidity that was quite marvelous as compared with the old method. In forming coffer-dams for piers and abutments of bridges, quays and harbors, and in piling the foundations of all kinds of masonry the steam pile driver was found of invaluable use by the engineer. At the first experiment made with the machine Mr. Nasmyth drove a 14-inch pile fifteen feet into hard ground at the rate of sixty-five blows per minute. The saving of time effected by this machine was very remarkable, the ratio being as 1 to 1,800; that is, a pile could be driven in four minutes that had before required a day. One of the peculiar features of the invention was that of employing the pile itself as the support of the steam hammer part of the apparatus while it was being driven, so that

FIG. 33.—WARRINGTON-NASMYTH HAMMER, FAIR HAVEN BRIDGE.

the pile had the percussive force of the deadweight of the hammer as well as the lively blows to induce it to sink into the ground. One of the most ingenious contrivances of the pile driver was the use of steam as a buffer in the upper part of the cylinder, which had the effect of a recoil spring and greatly enhanced the effect of the downward blow."

Many modifications of this hammer have been manufactured, and one much used at present is the Warrington-Nasmyth hammer, made by the Vulcan Iron Works. This hammer (Fig. 32) is made in three sizes, the weight of the striking parts being 550 pounds for sheet pile work, 3,000

FIG. 34.—CRAM-NASMYTH STEAM PILE HAMMER.

pounds for medium pile work, and 4,800 pounds for use on heavy work. This machine is provided with a positive valve-gear, a short steam passage to avoid the waste of steam, a wide exhaust opening to prevent back pressure as the hammer drops, a piston-head forged on the rod, and channel bars on the sides to allow the pile to be driven lower than the leads of the derrick. The hammer is attached to the hoist rope, but this is left slack when the hammer is resting on the head of the pile, steam is turned on and the hammer pounds automatically at the rate of sixty to seventy blows per minute until the pile is driven. The bottom casting which rests on the pile is a bonnet which encases the top and prevents brooming or splitting.

The hammer should have plenty of play in the leads, and the steam pipe should extend half way up the derrick to save length of hose. This

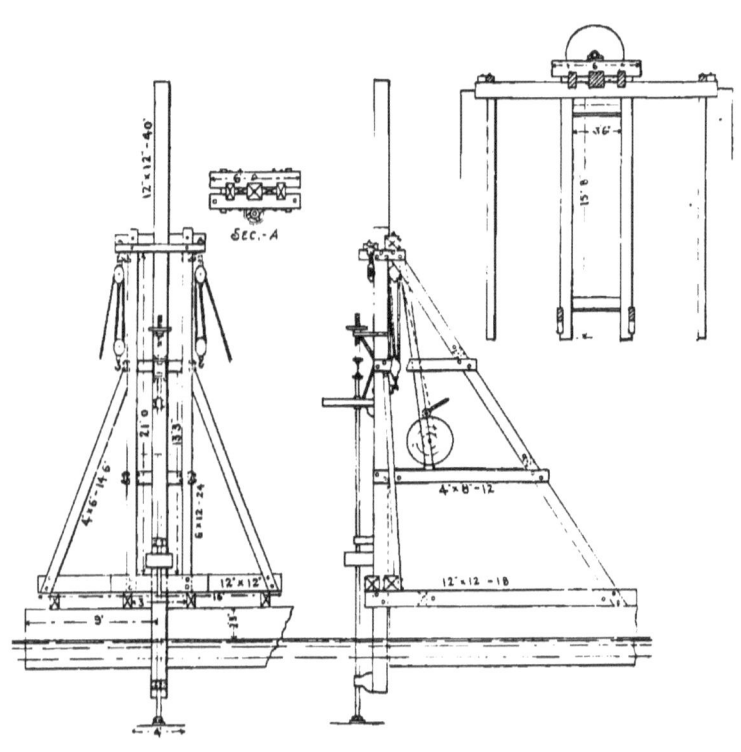

FIG. 35.—MACHINE FOR SAWING OFF PILES UNDER WATER.

hammer has a record of as high as seventy-five to one hundred piles per day, and one account gives the record of 3,000 lineal feet of piling per day at a cost of $50, the number of men employed being sixteen and the coal consumption one ton. This hammer is shown in Fig. 33 in use driving piles for bridge work on the Fair Haven bridge.

THE COFFER-DAM PROCESS FOR PIERS.

Another form of the Nasmyth hammer is the Cram (Fig. 34) which is very simple in construction. The driving head is hollow and the steam enters through a hollow piston rod, causing the head or cylinder to rise on the rod. Four sizes are made, with hammers of 430 pounds, 2,000 pounds, 3,000 pounds and 5,500 pounds. The small hammer which is listed at $300 is used for sheet pile work by inserting a "follower" of oak which fits the base or pile cap, and which has a slit in the lower end to fit the sheet pile. The number of blows per minute is the same as other steam pile hammers and an average of eighty-three piles per day of ten hours is reported, where they were driven seventeen feet into sand and oyster shells in the Passaic river, the largest day's work being 121 piles, or nearly double the best work with an ordinary hammer.

Mention has been made of the use of a rock drill as a Nasmyth hammer, on the Great Kanawah river coffer-dams; and where any amount of driving is to be done it will certainly be wise to use a hammer of the Nasmyth type.

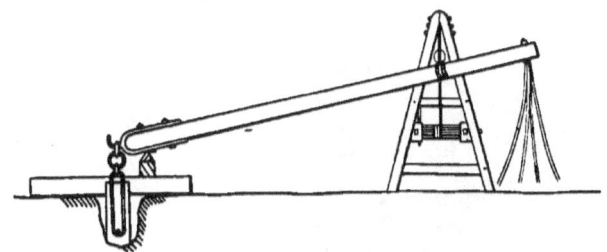

FIF. 36.—PILE-PULLING LEVER. AFTER CRESY.

The guide piles of a coffer-dam should always be driven with the idea of using them as a support for pumps, engines, derricks, and the like, although it will often be found cheaper to rig up on flat-boats when there is danger from floods. In determining what load a pile will carry from this source, or when driven as a foundation pile to support the masonry, Wellington's formula is at once the most accurate and the easiest to remember and use. For a drop-hammer, multiply twice the weight of the hammer in pounds by the drop in feet and divide by the last sinking in inches plus one, and the result is the load in pounds the pile will carry, with a factor of six for safety. This is easily remembered as 2 wh over s+1, and is always ready for use. For the steam-hammer the form is 2 wh over s+0.1, the "wh" representing the dynamic effect of the hammer.

Where piles have been firmly driven and they are to be removed when the work is done they can be cut off under the water by a machine similar to Fig. 35, which can be operated from a barge. The description in the *Engineering News* gives but little information in addition to the drawing. The shaft works in cast-iron sleeves attached to a timber, which slides in the leads, this being operated by the winch shown in side elevation. The final adjustment is made by the hand-wheel on the 3-feet adjusting screw. Where the piles are not so solidly driven they can be pulled out with a lever, an old form of which is given by Cresy (Fig. 36). In place of the pin and links, a chain closely wrapped around the top of the pile is usually made use of.

The apparatus used on the New York State canal work (Fig. 37) consisted of a strong frame mounted on a scow, from which was suspended a heavy set of falls to attach to the chain wrapped around the head of the pile. The pulling was done by an engine placed on the scow.

The construction of coffer-dams with sheet piling has led to the use of a

FIG. 37.—PILE-PULLING SCOW, NEW YORK STATE CANALS.

number of forms of sheet piles, some of which are driven only as a protection to the puddle, while others are nearly or quite watertight in themselves. The principal forms are shown in Fig. 38, the simplest form being plank of some considerable thickness (a) for which Stevenson specified 4½ inches by not exceeding 9 inches in width for the Hutcheson bridge. The points are sharpened as at (i) so they will draw together in driving, and as at (j) to cause them to drive straight and easy. The same principle is embodied in the patent metal point shown at (k), which is used to protect the point when driving through coarse gravel.

The piles at Buda-Pesth were increased to fifteen inches square in order to resist the pressure brought upon the sides of the dam by the puddle, the water, and also by the ice. Flat plank are also used by driving two or more rows as at (b), the second and third rows being used to close the cracks in the main row of piles and retain the puddle. An example of this will be given in the next article, where it was used on the Michigan Central Railway. The extra rows may be of thinner plank if they can be driven.

Mention has already been made, incidentally, of the use of V-shaped tongue and groove piling (c), on the Union Pacific Railway. This may be

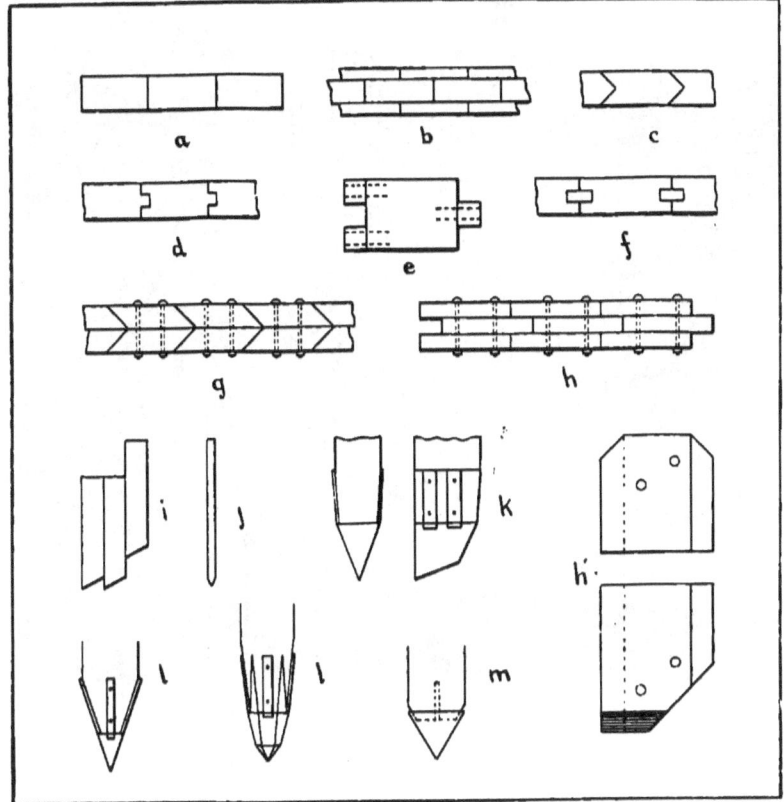

FIG. 38.—SHEET PILES AND SHEET PILE DETAILS.

made on a beveled saw table, the saw cutting half through the plank from opposite sides at each cut. This will produce a reasonably tight wall, if care is used in driving and if the points are sharpened to draw them together and make tight joints.

Ordinary tongue and groove piling (d) is frequently used, but a more frequent form is that shown at (e), like that used on the Robinson circular dam. The two pieces forming the groove and the piece for the tongue are

FIG. 39.—CHARLESTOWN BRIDGE. DRIVING WAKEFIELD SHEET PILING.

spiked to the 9x12 with 6-inch spikes sloping upward. A sheet pile dam on another pier of the Arthur Kill bridge, employed piling in which the grooves were made by making two saw cuts and cleaning out between with a chisel, the tongue being formed in the same manner as at (f), the tongue being spiked in one side.

A method which is not often employed is shown at (f) two grooves being made in the sheet pile and a key driven after the piles are down. Should the piles not drive in perfect line, and the groove fail to match, the method will not be found to be a success.

Sheet piling formed of two or more plank bolted together is being extensively used, one of them (g) being formed by two planks sawed with beveled edges and bolted together to form a pile similar to (c). This forms a pile which will drive easily on account of having some size and which will require fewer supports in the shape of waling pieces.

Several examples already given describe the use of Wakefield patent sheet piling (h), the method of sharpening being shown at (h'). This is constructed of three layers of plank from one to four inches thick, owing to the pressure to be sustained. The center plank must be sized to keep the tongue and groove uniform, and the plank are bolted together with six bolts for a length of from sixteen to twenty feet, two bolts near each end and two intermediate. For long piles, spikes should be driven between the bolts. The bolts vary from $\frac{3}{8}$ inch for 1-inch plank to $\frac{3}{4}$-inch for 4-inch plank. A coffer-dam constructed with this piling is shown in process of construction in Fig. 39, for the foundations of Charlestown bridge near Boston. A description of this will be given in the next article.

Pile shoes for use on round or square piles are shown at (l) and (m), (l) being patent forms. Straps of bar iron are used in many cases with success, for main piles, and sheet iron of $\frac{1}{8}$-inch thickness, bent to a "V" and spiked on, is often all that is necessary when shoes must be used on sheet piles.

The thickness of sheet piling should be sufficient to prevent the plank from bulging and should be calculated to stand a water pressure due to the depth, and for a span equal to the distance between the waling timbers or other supports. This would necessitate wales every six feet for 3-inch plank under five feet head, or wales every three feet for a 21-feet head. Plank $4\frac{1}{2}$ inches thick would require wales every seven feet under a 9-feet head, or every five feet for an 18-feet head. Timbers nine inches thick will carry nine feet under a 20-feet head, while the 15-inch timbers of the Buda-Pesth dam would carry twelve feet under a 33-feet head.

Good timber should always be employed if it can be procured, or if faulty stuff must be used, allowance must be made by using thicker piles and by placing the wales closer together.

ARTICLE V.

THE COFFER-DAM PROCESS FOR PIERS.*

CONSTRUCTION WITH SHEET PILES.

WATER pressure against the sides of a sheet pile coffer-dam is seldom provided for in an accurate manner, the thickness of the piling being usually decided upon from past experience, as is also the size and spacing of the guide piles and wales.

These are points where guess-work should be eliminated, as otherwise good coffer-dams are often seen, where the pressure has so bulged the plank as to cause leakage. While this may perhaps be corrected by additional bracing, simple calculations may easily be made to determine the size beforehand.

The pressure against a coffer-dam may act as at (a), Fig. 40, the sheet piling being in the condition of a beam fixed at one end and loaded with a gradually increasing weight, as shown by the dotted lines, due to the pressure of water or puddle at 62.4 pounds per cubic foot. Then the load on a width w of the wall is $124.8\,w\,d^2$ and the moment of the pressure is $83.2\,w\,d^3$. Taking the allowable unit stress on wet timber at 1,000 pounds per square inch, the thickness t of the sheet piling may be obtained from the formula

$$t = \sqrt{.496 d^3}$$

in which d is to be taken in feet and the resulting value of t will be the thickness in inches of the sheet piling.

This formula has been expressed in a graphic manner in diagram (d), Fig. 40, from which, knowing the depth of water $2d$, the thickness of piling may be read directly without calculation.

The addition of a strut, as at (b), Fig: 40, places the sheet piling in the condition of a beam supported at the upper end and fixed at the lower end, but for practical reasons, it is best to consider it as merely supported at both ends. The load will be the same as in the former case, $124.8\,w\,d^2$, but the

---

* The assumption that the pressure of puddle will be the same as water-pressure is made advisedly. It is true that *very* wet clay, approaching a fluid condition, will exert a much greater pressure, but it would then be useless as puddle. Dry clay would exert a pressure of less than half that due to water, so it has been assumed that *wet* clay or puddle would exert the same force as water. Should it exceed it for a short time no damage would be done, owing to the low unit-stress adopted.

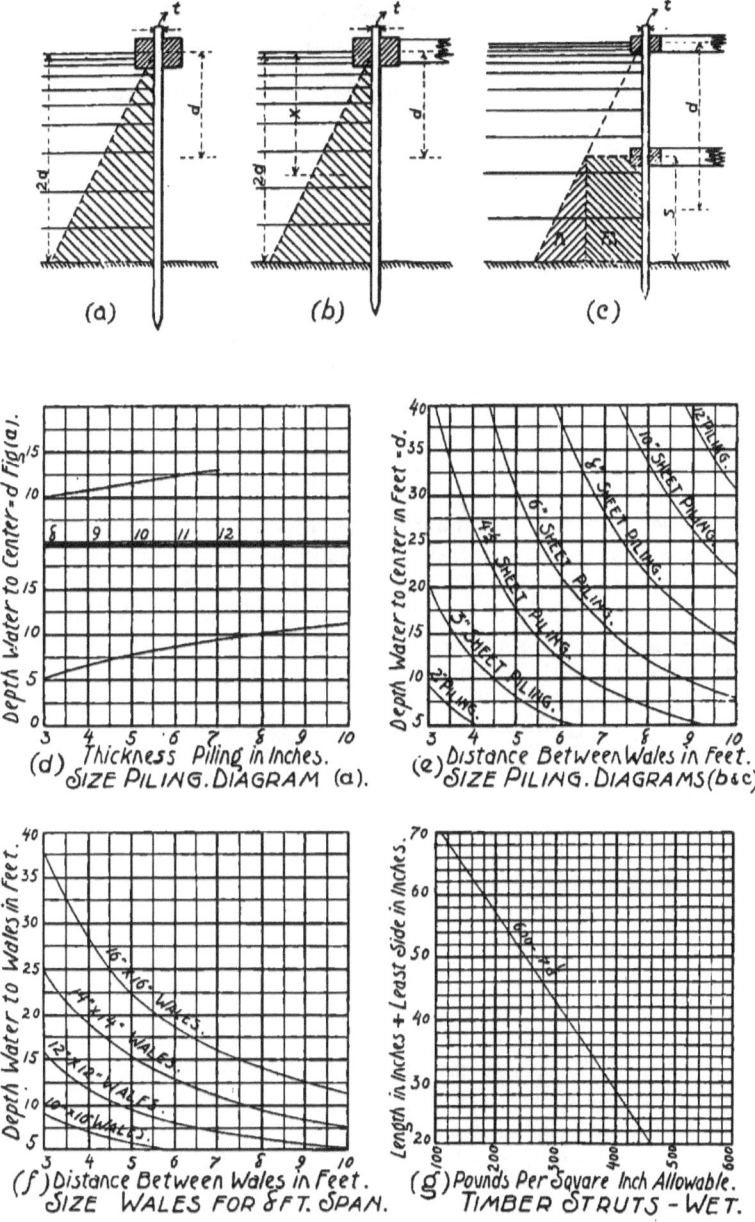

FIG. 40.—ARRANGEMENT AND DIAGRAMS OF SIZES FOR SHEET-PILE COFFER-DAMS.

maximum moment will occur at a point x which is a distance from the top equal to 1.16 times d, and has a value of $32\,w\,d^3$. The thickness t may be found from the formula

$$t = \sqrt{.192 d^3}$$

When the section of the plank to be calculated is located as "s" in (c) of Fig. 40, it is in the condition of a beam fixed at both ends and loaded with a uniform load m and a triangular load n. The exact analysis of this is too lengthy to be taken up here, and reference may be made to page 195 of "Wood's Resistance of Materials."

For practical purposes we may consider the load as all uniform and due to the head acting at the middle of the span. This will give a load of $62.4\,w\,d\,s$ on the span s for a width w, and a moment of $7.8\,d\,s^2$, which gives a formula for practical use, for a unit stress of 1,000 pounds per square inch of

$$t = \sqrt{.047\,d\,s^2}$$

This is closely represented graphically in diagram (e) of Fig. 40, which may also be used for case (b) by taking the depth of water to the middle of the span. For example, when the depth of water to the middle of the span is 15 feet, find this in the vertical column to the left, and if 6-inch sheet piles are to be used, follow the horizontal through 15 feet until it intersects the 6-inch curve and vertically beneath will be found the maximum spacing of wales, 7 feet 3 inches.

The size and spacing of wales may be taken from a similar diagram (f) of Fig. 40, which assumes the guide piles to be eight feet apart. The spacing of struts or braces will vary so much, that the load must be calculated, and when this and the length are known the size may be calculated from diagram (g) of Fig. 40, which is for wet timber.

From the formula

$$p = 600 - 7\,(l \div d),$$

in which p is the allowable stress in pounds per square inch, l is the unsupported length in inches and d the least side of the stick in inches.

Where two rows of sheet piling are to be driven to form a puddle chamber, if they are to be efficiently braced from the inside of the cofferdam, it will be sufficient to have a thickness of puddle of from two to four feet to exclude the water, depending on the quality of the puddle. Where there is to be no internal bracing, but two rows of sheet piling braced together together and filled with puddle are to resist overturning, the common rule is to make the width of the puddle chamber equal to the height above ground, up to 10 feet. When the height exceeds 10 feet, add one-third of the excess height to 10 feet for the width.

When the puddle chamber becomes very wide it is often divided into several compartments, as was shown in Fig. 5, and stepped in a similar manner. When the bottom is rock overlaid with a thin deposit of clay or gravel, the sheet piles may be driven around an open crib-work for support, as was done at Harper's Ferry, on the B. & O. R. R.

Where guide piles are to be used, the waling pieces are framed in, as was specified on the Hutcheson bridge, as shown at (a), Fig. 41, where the guide piles are of sawed timber. The wales are spaced slightly farther apart than the thickness of the sheet piles, to allow clearance in driving, the space between the guide piles being filled out with a key pile to fill the panel tightly. This method is but little used with tight piling, that shown at (b), Fig. 41, allowing the piling to be driven continuously, by removing the

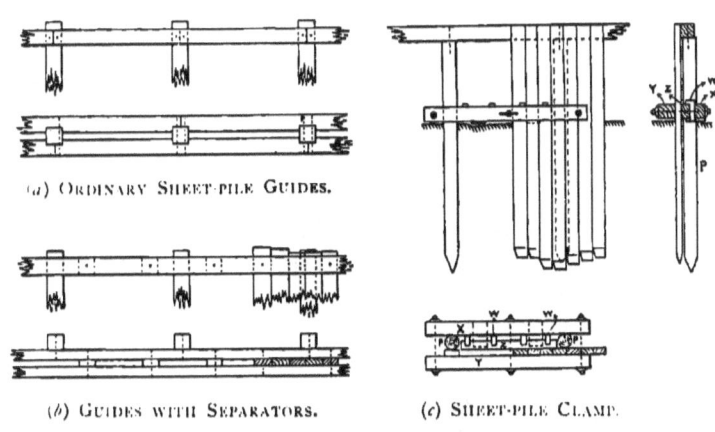

(a) ORDINARY SHEET-PILE GUIDES.

(b) GUIDES WITH SEPARATORS.   (c) SHEET-PILE CLAMP.

FIG. 41—SHEET PILE GUIDES AND CLAMPS.

spacing blocks as they are reached, and substituting bolts through the sheet piles, firmly connecting the piles and wales together,

A very satisfactorly method is described in the *Engineering News* of May 12, 1892, which was used by A. F. Walker. Having occasion to do a large amount of work it was desirable not to go to the expense of squared guide piles. Round guide piles (P) were first driven seven feet apart, and cut off to a level. Caps were then drift-bolted to the tops, allowing them to project slightly beyond the face of the round piles, thus forming a permanent support for the top of the sheet piles. Near the ground line was placed the clamp, consisting of two sticks (X) and (Y), connected by three bolts and drawn together as tight as the intervening piles or pile and gauge block (G) will permit. The stick (Z) is then forced forward by the wedges (W) until

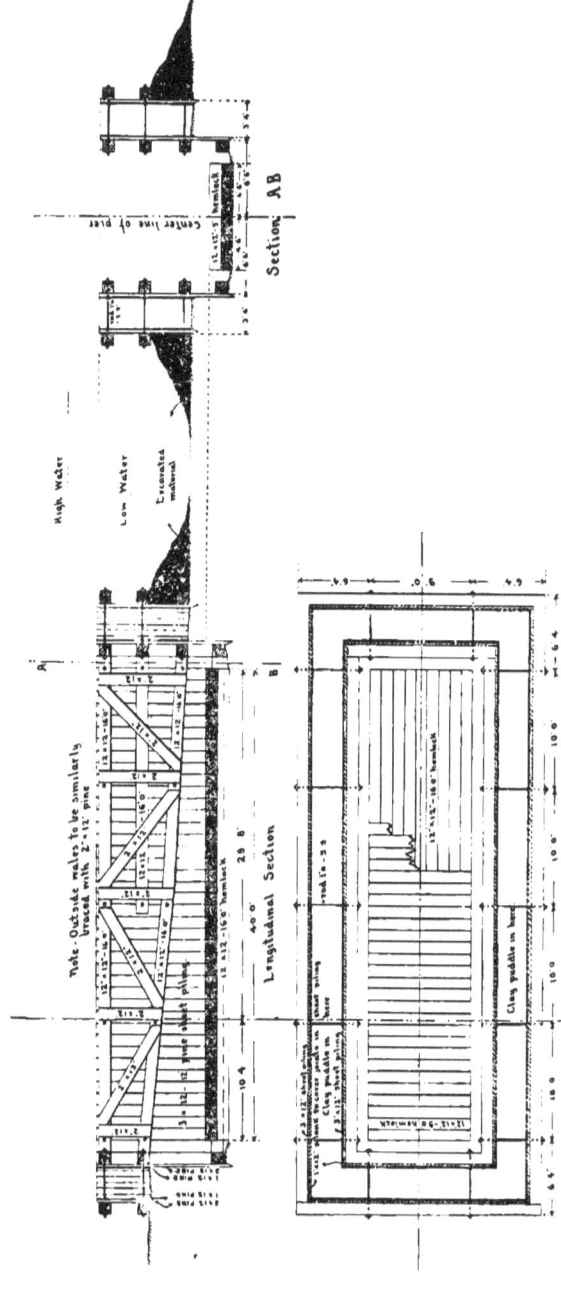

FIG. 42.—COFFER DAM FOR ANN ARBOR BRIDGE, MICHIGAN CENTRAL RAILWAY.

the space between (Z) and (Y) is the same as the thickness of the piles. The pieces (X) (Y) (Z) are slotted for the middle bolt, and this permits of some adjustment. When one of the piles partially closes this slot, a notch is cut in the same large enough to receive the bolt, and the bolt is then slipped up to it and tightened. This allows of the next pile being driven as close as the others. When one panel has been completed the nuts are removed and the clamps moved forward to the next one, a notch being cut in the end pile to receive the end bolt of the clamp. The piles are sharpened flatwise with a little more slope on the side facing the guide piles, giving them a tendency to drive away from the guide pile at the foot and bear against the cap at the top. A slight bevel is also given to the edge to make the foot crowd the adjoining pile. During the first half of the driving, the joint is held a little open at the top, but during the latter half, pressure

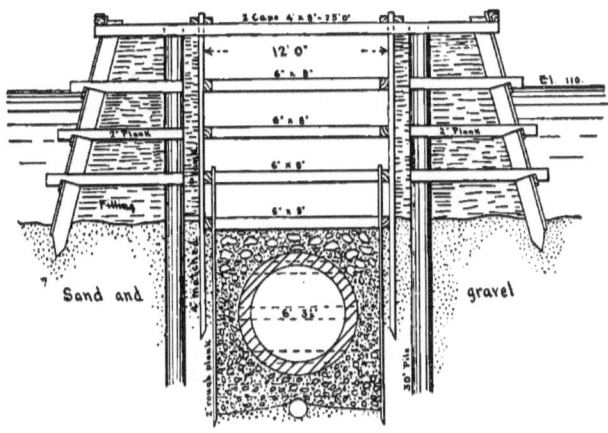

FIG. 43—SEWER COFFER-DAM. BOSTON SEWERAGE SYSTEM.

is brought to crowd it toward its neighbor, and the joint will close as tightly as possible.

The use of single pieces of timber as wales, against which the sheet piling is driven, is illustrated in the use of method (b) of Fig. 38, by Benj. Douglas, Bridge Engineer of the Michigan Central Railway. The coffer-dam (Fig. 42) was built without guide piles, the wales being 12x12-inch timber bolted against the outside of the sheet piling, by the brace rods one inch in diameter. The wales are held in place vertically by bracing of 2x12-inch pine plank, which are spiked on as verticals and diagonals to form a truss and also to stiffen the framework in general.

The sheet piling is 3x12, and after being driven into the hard gravel bottom, the cracks were lapped by 1-inch boards The bottom was uneven and accounts for the difference in height, the excavation at the high end being dumped outside at the low end, to assist in making the dam tight. The puddle chamber was 2 feet 8 inches wide and was filled with clayey gravel. The plan also shows the grillage in place for receiving the foundation courses of the stonework. This is formed by 12x12 timber crossed, and drift-bolted together with 1-inch round and 18-inch long drift bolts.

The account of the Arthur Kill bridge foundation in Vol. 27 of the "Transactions of the American Society of Civil Engineers," by A. P. Boller, Consulting Engineer, covers a very interesting experience with sheet piling on Pier No. 5: "This pier is near the edge of the marsh forming the Staten Island shore, which is barely flooded at extreme high tides. Borings indicated about thirty feet from the surface to hard bottom, consisting of mud, mud and clay, clay and shale to the bottom of shaley clay, in which the pier was to be founded. Experience on other work of a similar character, indicated that the founding of this pier would be accomplished with little difficulty. The area of the foundations was inclosed with a tongued and grooved sheet pile dam of 4-inch yellow pine plank. But it was found impossible to hold the plank at a depth of fifteen feet, the mud and clay becoming puddled with water, and despite all efforts at bracing, the plank shoved inward to such an extent, as to spoil the whole dam before we were half way down. A second dam was therefore driven around the first one, but this time with 10x12-inch tongued and grooved timbers, in one length to reach the extreme bottom. These timbers were grooved by slitting the grooves out at the mill with a circular saw and chiseling the blank so formed free. The tongue was an independent spline, 2½x4 inches, of dry wood and nailed in one groove. The timbers were shaped at the feet to drive close. This dam was hard driving, but was finally accomplished, when digging was resumed and the old dam removed piecemeal as we could get in the braces. The bottom was reached within a perfect dam, with only one bad leak in the northwest corner, due to the shattertng of a small piece of one tongue during the driving. As it was impossible to stop this leak from the inside, and the outside was inaccessible, to prevent washing the concrete, the leak was led off in a box at the side of the dam to the sump well, and the footing course of concrete, filling the whole area of the dam about seven feet deep, was gotten in place."

This example emphasizes in a very decided manner many of the statements that have been made heretofore. While no doubt the removal of the old dam was attended with much expense, its inclosure entirely within the new sheet piling rendered the prosecution of the work comparatively certain.

An example of the driving of sheet piling on a slant, to prevent crowding in at the bottom is shown in Fig. 43, which is a cross-section of a sewer coffer-dam used on the Metropolitan Sewerage Systems of Massachusetts by

Howard A. Carson, chief engineer, and described in the *Engineering News* of Feb. 8, 1894.

The outlet into the ocean at Deer Island begins at a point about sixty feet inside the high water line and about 1,850 lineal feet is from five to ten feet below high water. This necessitated the coffer-dam, which was constructed with bents every six feet and with 2-inch plank inside the high water line, but for the remaining distance of 4-inch matched plank. The excavation was done by means of buckets, traveling derricks and dump cars, the latter being emptied at the sides and ends of the trench. The leakage from the ocean was kept out by using centrifugal pumps, which pumped a maximum of 46,000 gallons per hour. The concrete, which has large boulders imbedded in its surface the size of paving stones, was carried up to the level of the ocean bottom.

From the middle of June, 1893, when the work was begun, to the end of September, 526 feet of trench was completed. The size of the trench was 14 feet average depth and 10.8 feet average width, which made the excavation average 5.6 yards per lineal foot. The cost for the trench, including coffer-dam, sheeting left in, and back filling was $44.00 per lineal foot.

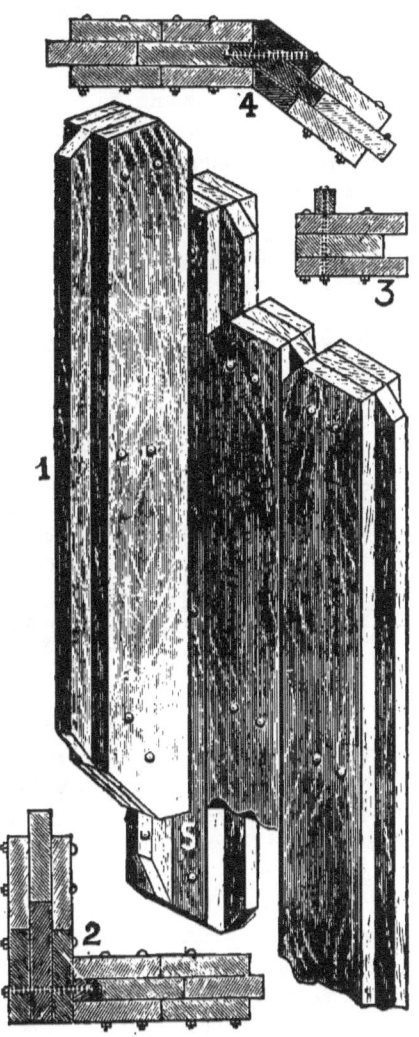

FIG. 44—WAKEFIELD SHEET PILING.

Casual mention has been made in several places of the use of Wakefield sheet piling which was illustrated at h and h' of Fig. 38 and which is further shown in Fig. 44. View No. 1 is of a corner which is formed as in the plan

No. 2, a tongue being bolted on the side of a pile, when the corner is reached as in No. 3. Any angle is turned by a similar method, which is shown by No. 4, or the piles may be driven to form a curve. The essential features of the system are the triple lap or long tongue and groove which excludes the water, and the use of ordinary plank, which can be easily obtained. The center planks should be sized to a uniform thickness, to insure the tongues fitting the grooves, and to make driving easy, while the three plank are to be bolted and spiked together to cause them to act as a compound beam and not as separate plank like the system of (b) Fig. 38. It is recommended to use a 2½-inch tongue on 1-inch boards and ⅜-inch bolts. For 1½-inch plank a 3-inch tongue, for 2-inch and 2½-inch plank a 3½-inch tongue and ½-inch bolts, while for 3-inch plank a 3½-inch tongue and ⅝-inch bolts are to be used, and the same size bolts for 4-inch plank, but a 4-inch tongue. Two bolts are to be staggered in every five to eight feet of the length of the pile and spikes used between the bolts on long piles.

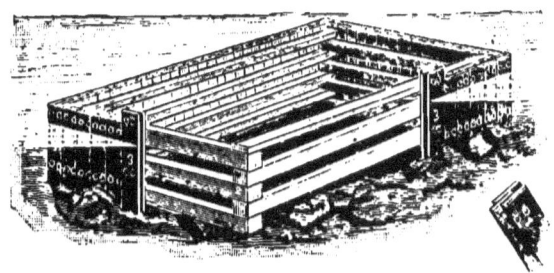

FIG. 45—TYPE OF MOMENCE AND HARPER'S FERRY COFFER-DAMS.

The La Grange lock on the Illinois river was inclosed with this piling, under the direction of Major W. L. Marshall, Corps of Engineers. It was intended to back the sheeting with earth, but as both dredges broke down the water tightness was entirely dependent on the Wakefield piling, and under a 7-feet head no leaks were developed. The piles were made of three plank 3x12 inches by 22 feet long and with a 3-inch tongue; they were driven by three pile drivers with hammers of from 2,800 to 3,000 pounds through sand and mud, and in one place a layer of shells. There was no difficulty experienced in driving the piles without special appliances.

The use of 1-inch boards in this form (Fig. 45) is described by H. F Baldwin, chief engineer of the C. & E. I. Railway: "In constructing our second track over the Kankakee river at Momence, Ill., it was necessary to extend the piers in that river. The bottom is limestone and the surface is very irregular. We tried several days and finally succeeded in constructing a coffer-dam with two parallel walls of sheet piling. We then tried the

Wakefield triple lap piling, constructed of 1-inch boards sharpened to an edge, 2½-tongue and groove, which were driven with sledges until the piles, which were soft pine, conformed to the uneven surface of the rock. This piling was driven around cribs loaded with stone, and after the piling was driven, gravel was put outside the coffer-dam, after which no trouble was experienced in pumping out the water."

The work on the foundations of the new B. & O. R. R. bridge, over the Potomac river at Harper's Ferry was similar in many respects to the above, and the system was found to be very satisfactory.

Reference was made to the use of this piling on the Charlestown bridge

FIG. 46—COFFER-DAM ON CHARLESTOWN BRIDGE.

at Boston and the driving of the piles shown in Fig. 39. The work was under the charge of Jno. E. Cheney, Consulting Engineer, and was successfully carried out. The piling were driven principally as forms for concrete foundations and but little care was taken to make the dams watertight. After the concrete was deposited they were used as coffer-dams against a 6 or 7-feet head of water. They were 18 feet 6 inches by 110 feet (Fig. 46) and in some cases were thirty feet below low water or forty feet below mean high water. The piling was made of 2-inch plank and driven with an ordinary pile driver. The pumping was done with a 20-inch centrifugal pump and in some cases a 12-inch Follansbee pump of the propeller type was used. The construction of the sewerage system at Fort Monroe, Va., under Capt. Thos. L. Casey, Corps of Engineers, is described in the report of the Chief of Engineers of 1896. The work was done on the general plans of Rudolph Hering, Consulting Sanitary Engineer. One of the special difficulties encountered "was the building of a sewage tank fifty feet in diameter, with walls of brick two feet in thickness, exteriorly diminishing to three feet at the center, the inferior reference of which was twenty feet below low water. As described in the report referred to, this was accomplished very successfully by excavating a large area to the reference of ground water, some five or six feet below the surface, and then driving by the pile driver and water jet combined, two concentric twelve-sided polygons of Wakefield sheet piling 28 feet in length, 30 and 22 feet from the center, about the circumference of the shallow excavation. (Fig. 47.) The material, consisting of fine water-soaked sand, with a small admixture of clayey matter and fine gravel, was then excavated between the polygons to a reference of 20 feet, transverse shoring braces bearing upon stout stringers being put in at intervals as the work proceeded. The material did not vary much in its general nature, but a number of old piles were taken up, some of which did considerable injury to the sheet piling when driven, as shown in the subsequent excavation. The water was controlled by a powerful steam pump having its point of suction fixed, the water being permitted to flow toward it throughout the circumference. It was noticed that ground water came through the sheeting very freely at first, but that it constantly ceased to flow to any great extent at a height of a few feet above the point of excavation as this continually descended, owing to the rapid drainage of the strata. The interior core, in fact, became quite dry, so that in excavating after the walls were laid, no water was encountered until the bottom of the external concrete ring had been virtually laid bare. Upon attaining the reference—20 feet, the excavation ceased and hand-mixed concrete was deposited directly upon the bottom, as this was considered to be sufficiently firm, the pump being stopped temporarily in order to prevent a flow. The concrete was rammed firmly against the outer sheeting externally and against plank forms with triangular cross-section resting against the inner sheeting internally, until six feet in depth had been put in place. The portion of the ring at the pump suction was filled rapidly with concrete in bags. The 2-feet brick wall was then carried up from the axial line of the concrete ring, the space between the wall and the outer sheeting filled with sand, except about six inches at the base of the wall, which was of concrete. The braces were removed as successively attained, the inner prism of dry sand being held securely by the sheeting and the extreme top struts, which were left in place until the inner core was completely excavated. On the completion of the latter work to reference—20 feet, the water which came in freely from without under the concrete ring at several points was conducted in a peripheral trench to the fixed point of pumping. No water came upward and the middle portions of the bottom became perfectly dry. The inner sheeting was cut off at the base of the ring, boards were placed transversely over the peripheral trench, a duck tarpaulin coated with hot asphalt laid down, and concrete rammed in place until the concave bottom with sump channel had been completed, leaving only the pipe, through which the ground water had been pumped continually, night and day at about 1,000 gallons per minute, penetrating the concrete. In order to fill this pipe, it was cut off above the level of permanent ground water, and after the water

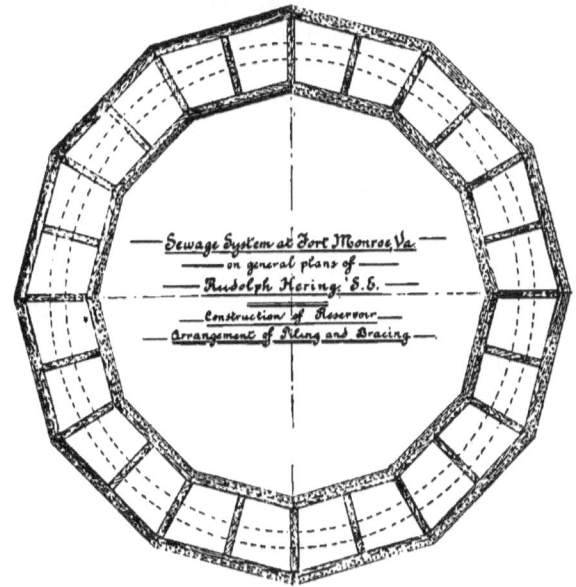

FIG. 47.—RESERVOIR COFFER-DAM. FORT MONROE, VA.

within had attained the level of ground water in the surrounding area and had become perfectly quiescent, neat cement in paper bags was dropped within, being retained at the bottom by the closed valve; the bags were readily broken up by a long pole thrust down the pipe. The latter was then cut off at the level of the bottom and a coating of cement plaster applied throughout. The resultant leakage through the bottom did not exceed about a gallon a minute and this will be greatly reduced by the infiltration of sand from beneath."

Further illustrations of the use of sheet pile coffer-dams will be given; then the operations of dredging, pumping and concreting described at some length.

ARTICLE VI.

THE COFFER-DAM PROCESS FOR PIERS.

CONSTRUCTION WITH SHEET PILES.

VARIOUS combinations of the sheet piling shown in Fig. 38 may be made, when occasion demands, or modifications may be made that will perhaps render the available material more effective. For example, the form (g) may be modified to the form shown in Fig. 48, which has the advantage of a wider lap, and should the piles not draw tight together in driving, no crack will be left open to admit the water. Then the piles of this form will act as guides to the ones being driven, similar to the ordinary tongue and groove piling. Other combinations and arrangements will readily suggest themselves as necessity may demand.

The use of sheet piling is often accompanied by a great deal of trouble in producing tightness, and as a matter of precaution, the very best method possible should be adopted in making the piling.

The coffer-dams constructed at Chattanooga for the Walnut street bridge over the Tennessee River, under Edwin Thacher, Consulting Engineer, were described in the *Engineering News* of May 16, 1891.

Four piers were founded by this method, but the account of pier number two will fully illustrate the work. The bed rock which was level, was covered by cemented sand, gravel and boulders, of which 320 yards were removed. The coffer-dam was built eighteen feet high, or eight feet above low water, to provide for a future rise. The inside was made large enough to allow of a space of four feet all around the base of the pier, and the space between the sheet piles for a puddle chamber was made nine feet. This was filled to an average of twelve feet with a clay puddle, of which there was 900 yards used. As a protection, there was placed outside the dam about 450 yards of puddle, and a breakwater was built up stream. About 38,000 feet of timber was used in the dam and breakwater.

After the dam was completed a rise of thirty feet washed out about half the puddle, and one end was crushed by a raft, but the repairs were made without serious trouble. No extra amount of pumping was required on any of this work except pier number three, where the seams in the bed rock required pumps with a capacity of 5,000 gallons per minute, and these did not suffice to keep the water down, until the seams were closed by laying sacks of concrete over them and weighting them down with large stones. The location of these seams is shown in Fig. 49.

The framework and wales for a sheet pile coffer-dam, used in founding the pier for the Baltimore street bridge at Cumberland, Md., are shown in Fig. 50, and this was described in the *Engineering News* of July 21, 1892, by H. P. Le Fevre, engineer in charge. The frame was built in place on two canal-boats and after completion was suspended from the old Bollman truss which the new bridge replaced.

The depth of the water was four feet, and about six feet of very loose quicksand and small round pebbles overlaid the hard bottom.

After the boats were removed, the frame was lowered to its place, the sheet piling driven and the dam pumped out with a six-inch pump. The foundation was laid on the hard bottom under the quicksand, after this had been removed.

The grillage was made of two courses of 15x15-inch clear white oak, around which was built a framework, and the open spaces of the grillage were then filled with a concrete, made up of one part of Cedar Cliff cement to two parts of sand and four parts of hydraulic limestone, broken to pass

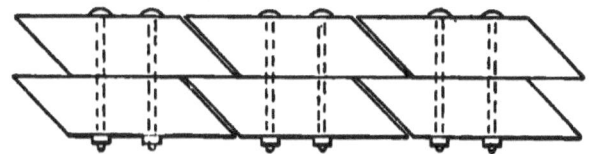

FIG. 48—COMPOUND SHEET PILE.

through a two-inch ring. Upon this was laid the footing courses of the masonry.

Another ordinary sheet pile coffer-dam which gave good satisfaction, was used at the Sandy Lake dam on the Mississippi River, by Major W. A. Jones, corps of engineers, and as the account contains so much of value, it will be quoted in full from the 1894 report of the Chief of Engineers.

"The coffer-dam is composed of two rows of round piles, twelve feet from center to center of piles, with the exception of sixty-two feet of the east end of the upper part, where they were driven fourteen feet from center. The piles in each row are eight and one-half feet from center to center, cut off at an elevation of 1217 feet above sea level and capped with 12x12 inch timber. The inside row of sheeting is 4x12 inch, and the outside 6x12 inch plank. The sheeting is cut off at an elevation of 1218 feet above sea level, or two feet below the flowage line. One-inch rods of round iron, eight and one-half feet apart, pass through the caps to prevent the filling from spreading the two lines of sheeting at the top.

In May, 1892, when a flood occurred, the outside of the cofferdam was raised three feet by splicing three-inch planks to the outside row of sheeting and then filling the triangular prism thus formed with earth. The cross section of Fig. 51 gives an idea of the dam above the bottom, while the longitudinal section shows the framing down to where it rests on the bottom, the frames being joined by the one-inch lateral rods of iron.

The total length of the coffer-dam is 829 feet, of which 742 feet is like that shown in cross section and the other 87 feet like that shown in the longitudinal section.

The number of round piles driven in the foundation is 1,605. The driving was commenced on November 12, 1891, and completed on August 21, 1893.

The material in the foundation is sand, excepting in the lower right hand corner, where there is some blue clay overlying the sand. The sand in the foundation is not as compact as it is usually found in the bed of streams. In the south half of the dam, the surface settled from four to six

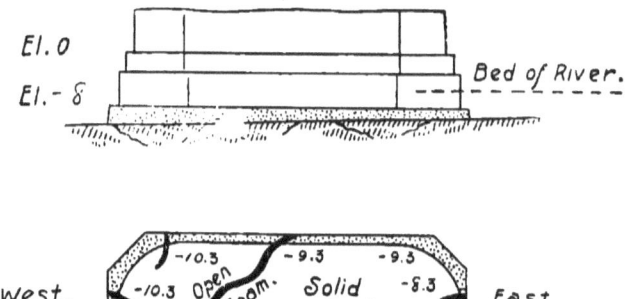

FIG. 49—CHATTANOOGA BRIDGE BED ROCK PIER NO. 3.

inches during the driving. As the surface was settling, the driving became harder all the time. In the north half, which embraces the navigable pass, there was some settlement, but it was not as noticeable as in the south half. The surface had probably settled by the jarring of the hammers while the first half was being driven. The penetration of the piles is also greater than it usually is in sand foundations in the bed of streams.

The piles were all of Norway pine and well seasoned. Two Mundy steam hoisting engines were used in driving, one a single cylinder and the other a double cylinder engine. In operating the hammer an inch and a half manila rope was attached to the pin connecting the lugs of the hammer, then passed over the sheave at the top of the leaders, and next around the drum of the hoisting engine.

When the hammer falls, it pulls the rope with it and unwinds it from the drum. This is what is termed driving with a "slack line." The blows are

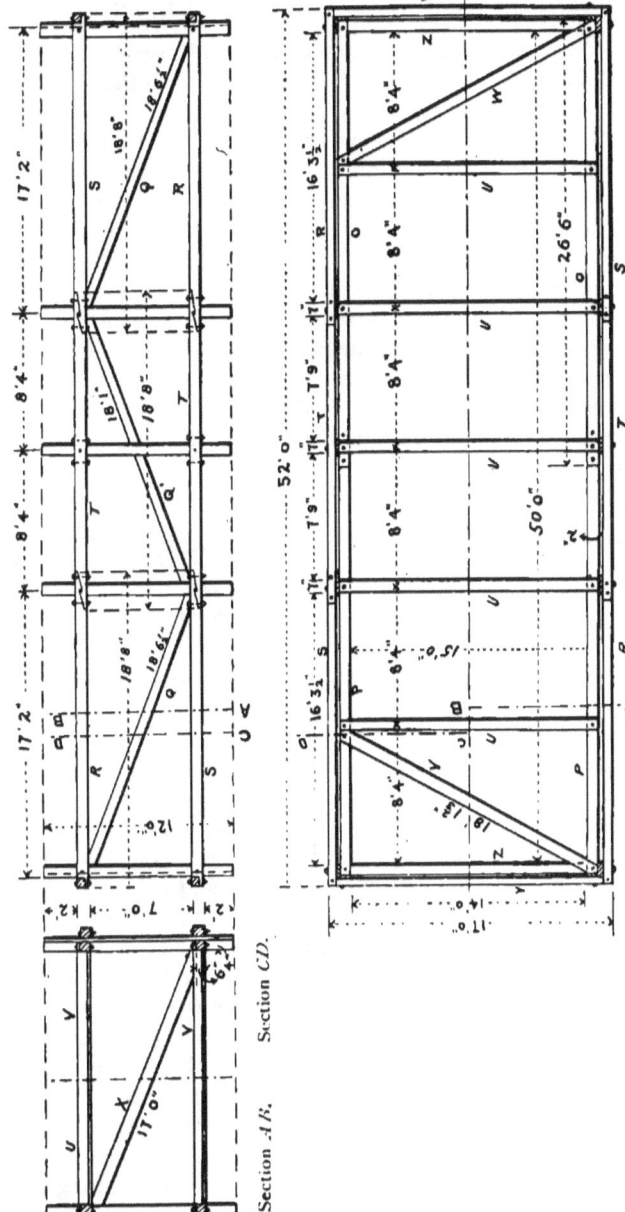

FIG. 50.—FRAMEWORK OF COFFER-DAM, CUMBERLAND, MD.

more rapid and keeps the material around the piles looser than it would be in the case of using nippers. Iron rings of ⅝x2½ inches Norway iron were used to protect the head of the pile.

It is a well-known fact in pile driving that it is very important to keep the material from settling around the pile, once it has been loosened, until the pile is down; for when the material has settled, or even partially, the penetration is diminished. The greatest load on a bearing pile is about 13½ tons.

Sheet piling was driven by a pile driver, assisted by a jet of water from a steam force pump. In driving all sheet piles a cast-iron cap or follower was used which fitted over the head of the pile. On the upper side of the follower there is a wooden block of some seasoned or close grained wood which receives the blow of the hammer. This device saves the head of the pile from being battered or splintered, and the pile can be driven to a greater depth than it could be without it.

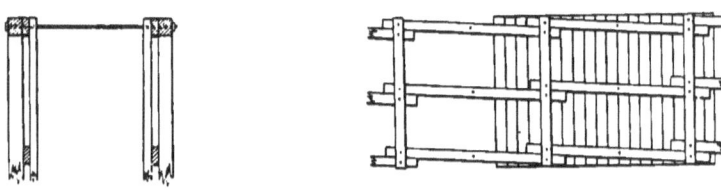

FIG. 51—SANDY LAKE COFFER-DAM.

In first using the jet on a sheet pile, a groove was made in the inner edge to receive a half-inch gas pipe, which was connected to the force pump by means of an inch and a half hose. The aperture at the lower end of the gas pipe was reduced to a diameter of about three-eighths inch. The water was thus forced to the bottom of the pile, and the sand loosened.

This worked well until the sheet pile struck gravel, when the nozzle of the pipe would become battered or filled with gravel. The pressure in the hose would then burst a coupling somewhere. Another source of trouble was the frequent breakages in the connection between the pipe and the hose, on account of the jarring of the hammer. This plan after awhile was abandoned and the nozzle of the pipe was thrust by hand under the point of the pile. The piles are driven in the ground from 12 to 14 feet.

The construction of the Main street bridge at Little Rock, Arkansas, involved the construction of two coffer-dams, for piers No. 5 and No. 6. This work was done under the direction of Edwin Thacher, Consulting Engineer, whose original specifications called for pile foundations for these piers, the piles to be driven to bed rock and cut off four feet below low water, to receive a grillage of 12x12-inch timbers to receive the masonry. The size of

FIG. 52—COFFER-DAM AND CONCRETE PIER, LITTLE ROCK, ARK.

the grillage being 12 and 13 feet wide by 34 feet long and resting on forty-eight and sixty piles respectively, the piles being of good sound oak or pine at least eight inches in size at the small end and not less than twelve inches at the butt when sawed off.

The coffer-dams were constructed, as can be seen from the view in Fig. 52, by driving guide piles, to the top of which are drift bolted square guide timbers. The sheet piling of three-inch tongue and groove stuff was driven against the outside of this timber, and the excavation banked up against the outside. They gave excellent satisfaction and caused little trouble as the water was shallow.

The piers were constructed of Portland cement concrete, the facing of two inches thickness being a mortar of one part of cement to two parts of sand while the balance was of concrete of one part cement, three parts sand and six parts of broken stone.

Where sheet piles are to be driven on rock bottom or through earth or gravel to rock bottom, they should be driven hard enough to broom up and form a close joint with the rock. This has been accomplished also by driving the piles with a thin edge until they fit the rock bottom, when they are drawn and after cutting them to conform to the contour of the rock, they are redriven, thus forming a tight joint. This method while very good, is too expensive for general adoption.

Coffer-dams are quite frequently constructed for the repair or removal of existing piers. A pier which was constructed in 1840 in the river Parnitz, at Stettin, Germany, became an obstruction to navigation and it was decided to remove it. The work was described in the *Engineering News* of July 14, 1892.

Its exterior showed a facing of granite laid in hard Roman cement, and soundings revealed the existence of a course of sheet piling around the pier, with a protection of rip-rap at its foot. The original drawing of the pier showed a pile foundation. The specification prescribed the use of the old course of sheet piling, shown at A on accompanying cuts, for the construction of the coffer-dam. Owing to the belief that the existing sheet piling, after having served such a length of time, would not be sound enough to permit of its use in the erection of a coffer-dam, local contractors could not be found and the work was let to an outside contractor.

The preliminary work was begun by picking up the rip-rap around the foot of the pier with a claw dredger mounted on a raft. Some of the stones weighed as much as a ton. The bottom of the river, after the rip-rap had been cleared away, was found to be covered with a layer of concrete, consisting of pieces of brick and cement. This was brought up in large slabs. The pier itself was found to be of rubble masonry, composed of irregular shaped granite blocks with the interstices filled with brick, laid in cement mortar. The single stones were detached and swung off by the claws of the dredger. Their average weight was about one and a half tons.

After the masonry had been pulled down to nearly the level of the water a row of sheet piling, shown at b in Fig. 53, consisting of piles seven inches thick, was driven to a depth of nearly ten feet. The space between the old and new sheet piling was filled with blue clay. To keep the interior free from water two pumps were employed. After putting in the necessary bracing the work of removing the masonry to the bed of the river was continued. A shell of the latter, however, was left standing. Then the timber platform on which the masonry had been resting and the layer of concrete below were taken out, exposing a layer of clay underneath. While attempting to pull one of the foundation piles a stream of water rushed through the opening thus formed, so that this plan had to be given up and blasting re-

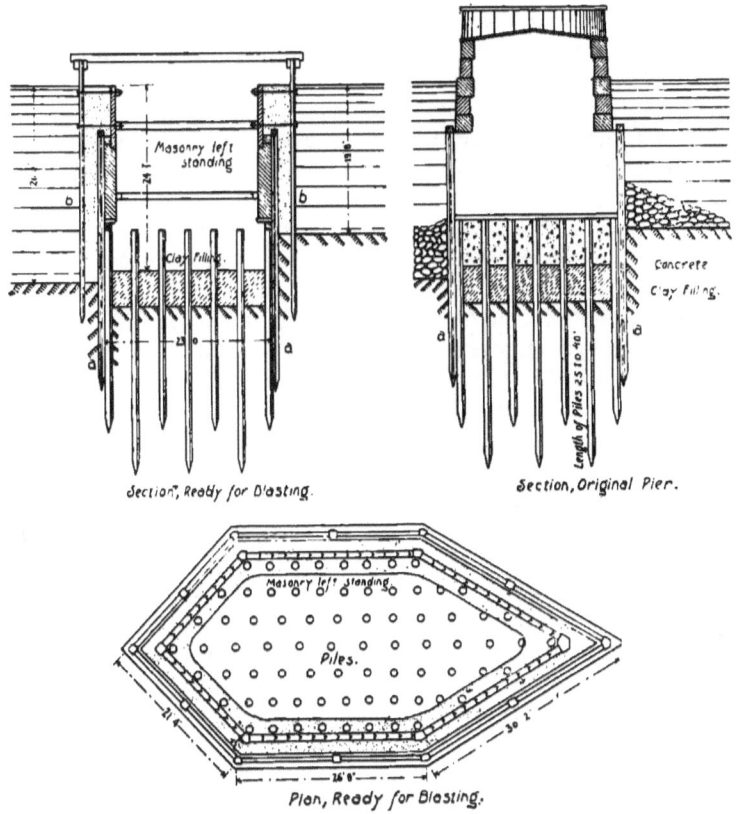

FIG. 53—REMOVAL OF MASONRY PIER AT STETTIN, GERMANY.

sorted to. To do this the tops of the piles were bored to a depth of thirteen feet and filled with 8.8 pounds of dynamite each. The initial charges consisted of 10.6 ounces in air-tight canisters. The shell of masonry left standing received four cubical charges of 8.8 pounds each. In all sixty-eight charges, consisting of 616 pounds of dynamite, were used. The electric current for the blast was divided into three currents, each being attached to an induction apparatus. The blasting, however, did not prove to be as effective as was anticipated, owing to the dissolving action of the water, and several charges were taken out intact. The clearing away of the wreck was almost entirely done by the claw dredges. The piles, which were split and loosened in their sockets by the force of the explosion, were pulled up by windlasses mounted on flat boats. The work of removing the pier lasted nearly nine months and the cost was about $8,700.

Another example of the removal of a pier was at Gadsden, Alabama, where a pivot pier in the Coosa river had tilted. The pier had been built originally in a water-tight caisson and was supposed to have been founded on solid rock, but by some error a layer of gravel was left underneath and eventually the pier tilted down stream seven feet, nearly throwing the swing span into the river.

After the span had been blocked up to allow the passage of trains, a cofferdam was built around the pier to give plenty of clearance to the old caisson. (Fig. 54.) This was constructed by driving three rows of sheet piling through sand and gravel to bed rock and puddling between them.

The sand and gravel over the rock was not removed from the bottom of the puddle chamber before puddling and a great deal of trouble was experienced all through the work by leakage through the porous gravel. It is probable, too, that a poor joint was made between the sheet piling and the rock.

Bents were erected upon the sides of the coffer-dam and by driving piles into the puddle and inside the dam, to carry a truss on each side of the span, which carried the drum and supported the main trusses at the center. When this had been tested by loading with trains of ore upon the bridge and found to be satisfactory, work was at once begun upon the removal of the old pier, by means of two fixed derricks on the false work and one floating derrick. The stones were marked as they were removed to insure their return to proper places when the pier was rebuilt, and were taken to the shore until needed again. When the masonry was all removed the grillage was broken up and taken out, after which the gravel inside the coffer-dam was cleaned out down to bed rock. New footing courses were laid to take the place of the gravel and old grillage, and the old stonework relaid by placing each course in its former position as nearly as possible. The pier was about 80 feet high and contained about 1,100 yards of masonry. The work occupied from Sept. 15 to Dec. 25, 1888, and was done under the direction of Cecil Frazer. The description is taken from the *Engineering News* of April 13, 1893.

The construction of the piers for the Philadelphia and Reading railroad bridge over the Schuylkill, was accomplished by the use of a floating cofferdam, the foundations being laid upon the bed rock.

THE COFFER-DAM PROCESS FOR PIERS. 75

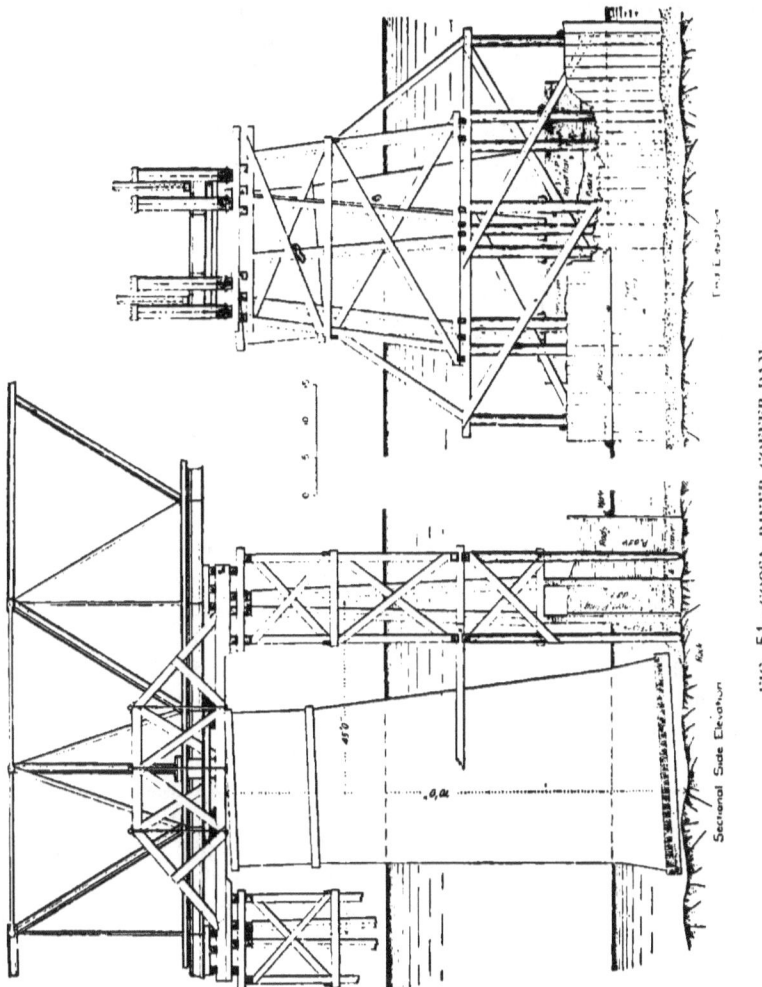

FIG. 54—COOSA RIVER COFFER-DAM.

When in position for work the dam is rectangular in shape, 62 feet long and 36 feet wide, outside dimensions, and 16 feet high. Each side consists of timber crib work 10 feet wide, making the inside dimensions 42x16 feet. At each corner there is a movable timber extending vertically from the bottom of the crib to some distance above the top. These timbers or spuds are shod with iron on the bottom, and serve to hold the dam in position while the sheet piling is being driven. The dam is divided vertically through each short side into two equal parts, which can be floated separately to any desired position and afterwards joined together. Watertight compartments are built in each section to assist in floating it, and these compartments are also used to hold stone when it is desired to sink the cribs.

When the two sections are united and placed in required position the spuds are dropped and the crib work is sunk by letting water into the watertight compartments, and putting in the necessary amount of stone.

Any irregularity in bearing between the bottom rock and the bottom of the crib is then corrected by a diver, who blocks up where required. Close sheet piling of jointed plank three or four inches thick is then put on the outside and spiked to the cribs. Puddle, composed of clay and gravel, is then thrown around the bottom outside, and the dam is ready to be pumped out. When the masonry reached the height of the braces they were taken out and the dam was braced against the masonry.

The maximum depth of water encountered at Falls bridge was thirteen feet at ordinary water level. Several freshets occurred during the progress of the work which did some damage to the dam. At one time, when a dam was ready to be pumped out, a rise in the river moved it down stream about thirty feet, tearing off the sheet piling. It was drawn back to place and successfully completed. To make a complete shift of the dam from one pier to the next, with a gang of six men, required about six or eight days, divided as follows: To take the dam apart and reset it, about three days; to sheet pile, about two days; to puddle, about one day; and pumping out and puddling meanwhile required about one to two days, depending on the amount of the leakage. At each shift, a portion of the plank sheet piling, perhaps 10 per cent, had to be replaced by new stuff. The pump used was located on a small steamboat, and was run by a steam engine. The amount of pumping required after the dam was once pumped out varied for the different piers; some dams required little pumping and others a good deal. Only one of the foundations required much leveling off of the river bed, and this one also gave considerable trouble to keep the water out, but the leaks were finally stopped by using gunny bags around them; the bags being drawn into the crevices by the force of the water, thus holding the puddle.

The floating dam was used for the three piers in the river channel, the two piers near the shore being put in with ordinary dams. The floating dam is still in good condition and could be used again if needed. The original dam of which the one used at the Falls bridge is an enlarged copy, was used for twenty-three or twenty-four settings.

The foregoing account is taken from the *Engineering News* of May 24, 1894, the description being by W. B. Riegner, who states also that the cost of the coffer-dam, including one set of sheet piling, was $3,000, while the total cost for five coffer-dams, including the two crib coffer-dams at the sides of the river, was $14,000.

The subject of subaqueous foundations has been very fully treated of in a series of lectures by W. R. Kinipple, M. Inst. C. E., before the Royal Engineers' Institute at Chatham, England.

The use of six-inch pitch pine close sheeting was made use of by him, for a quay wall in the harbor of St. Helier, Jersey. They were driven to rock or as deep as possible with a 2,800-pound hammer, and the tops cut off a few feet beneath half tide level, and clayey material banked up against the outside. The bottom through which the sheet piles were driven was sand and clay.

The rock was laid bare to a depth of as much as thirteen feet below low water and in sections which contained about 900 tons of water to be pumped out; this was done with a sixteen-inch centrifugal pump in about forty-two minutes.

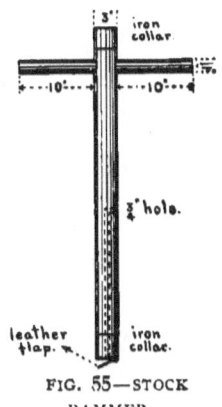

FIG. 55—STOCK RAMMER.

Several leaks were developed under the piles, but they were promptly stopped by "stock ramming." The stock rammer which is shown in Fig. 55, is 3 inches in diameter, 3½ feet long and banded top and bottom with iron. A ¾-inch air hole is bored up from its foot a distance of twenty to thirty inches, and covered on the bottom with a sole leather flap, so that air is let in and suction prevented as it is withdrawn. The sheet piles have 3¼-inch holes bored through their sides, and cylinders of clay are inserted 3x9 inches long, similar to the work at Sault Ste. Marie. The stock rammer is inserted and driven by mauls as far as its length will permit when it is drawn out, and other charges inserted until no more clay can be driven. The hole in the pile being filled with a wooden plug.

The piers for the Putney bridge, over the Thames, were built by the same engineer, with single pile dams to a great depth, by using fourteen-inch square piles, with elm wood tongues, and driving them down through the mud and clay to the stiff clay bottom, so that practically watertight work was secured

In the construction of the docks at Victoria, British Columbia, he constructed a coffer-dam 500 feet in length, in a depth of thirty-five feet of water, the bottom being of rock and overlaid in places with sand and shells several feet in thickness. At the center the sand and shells overlaid a bed of clay.

Three rows of close 12x12-inch sheet piling were driven with two puddle chambers of seven feet each between. The guide piles were 15x15 inches and the wales were 12x12 inches.

Where the dam rested on rock at the ends, heavy shoes were used on the piles and concrete deposited around their feet to make the work watertight. The dam was completed in October, 1879, and remained thoroughly tight until the dock was completed over seven years later.

The arch bridge at Topeka, Kansas, over the Kaw river, which is being constructed on the Melan system, of concrete and steel, by Keepers and Thacher, the designing engineers, is a most interesting piece of work. The coffer-dams were required by the specifications to be watertight, and to

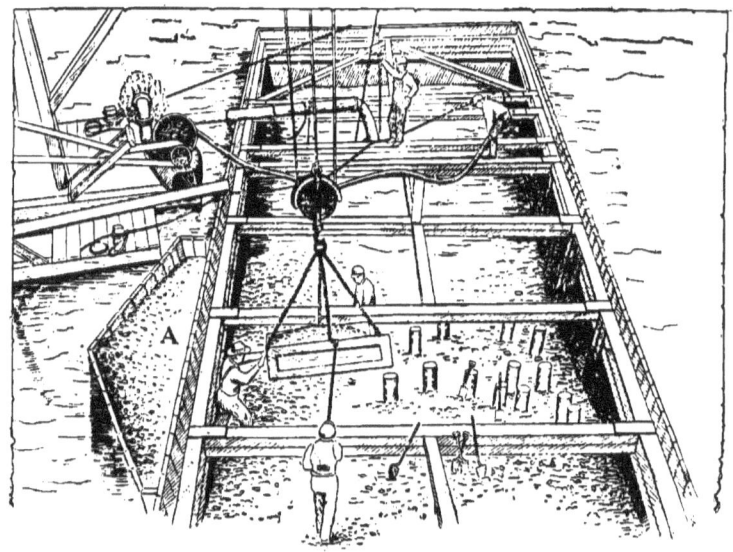

FIG. 56—TOPEKA BRIDGE COFFER-DAM NO. 4.—"A" shows puddle to stop leak.

effect this 4x12 inch tongue and groove sheet piling was used. The size of the coffer-dam for pier No. 4 was 18x55 feet in the clear (Fig. 56) and the piling was driven about sixteen feet into the sand bottom or twenty-two feet below low water. The driving was done by a 1,600-pound hammer with thirty-six feet leads; the power being furnished by a 15 H. P. hoisting engine.

No puddle was used around the outside except to stop leaks, and the dam was kept clear of water with a No. 6 Special Van Wie sand pump. The capacity of the pump was 3,000 gallons per minute of water, and from sixty to eighty yards of sand per hour. It was operated with a 15 H. P.

engine. The other piers were handled in a similar manner and with no particular trouble.

The growing scarcity of timber will doubless lead to the use of metal at some time in the future, to replace sheet piling for coffer-dams, but where timber is abundant and reasonable care is exercised in its use, it will continue to be of great service in obtaining foundations by this method.

ARTICLE VII.

THE COFFER-DAM PROCESS FOR PIERS.*

METAL CONSTRUCTION.

THIN steel shells have been used extensively for foundation work, but in the majority of cases they have been retained as essential features of the permanent construction.

This is more particularly the case in locations where stone is scarce or expensive and it becomes necessary to substitute some other material for foundations. Tubular steel piers are constructed of two tubes, ranging from 24 inches to several feet in diameter, or in the case of pivot piers, from 15 feet, with a single tube for a pier, to 30 feet or more.

In a number of instances the steel shells for ordinary piers have been made oblong, in the general form of a stone pier, and braced internally to hold them in shape during sinking, after which they are filled with concrete.

The metal shells for the Hawkesbury bridge in Australia were of this character, 20 feet wide, 48 feet long and with rounded ends. Each one was provided with three dredging wells, each 8 feet in diameter, through which the dredges shown in the view (Fig. 57) were operated. While these piers were not used as coffer-dams, they were made water-tight by boiler riveting, so that by pumping water in and out the displacement could be kept constant, and in this way control the pier in an average tide of five feet. These piers were sunk, by dredging out the material from the inside, to the great depth of from 135 feet 8 inches to 197 feet below the pier tops, or a distance of 155 feet below low water.

Both inclined and vertical cutting edges were used, with the result that

---
* Metal caissons have been used much more frequently in this country than have metal coffer-dams, the reason being the cheapness of timber and its more easy application.

In England metal coffer-dams are more frequently used. The example given in this article—the Forth bridge coffer-dams—might have been supplemented by reference to those used on the Clarence bridge at Cardiff, the construction used being illustrated and described in *Engineering*, and is especially notable for the design of the bracing.

the inclined ones were of frequent trouble and the vertical ones none whatever.

"If it is essential to increase the bearing surface at the bottom of the caisson to an area which is not required in the upper portion, this end can be secured by a vertical cutting shoe of considerable height, with a step or steps into the smaller diameter. This is quite as efficient to secure the end in view as a long incline on the cutting shoe, and has decided advantages. In the denser material the vertical sides leave the ground undisturbed for some height close to the skin of the caisson, and a vertical guide is secured

FIG. 57.—HAWKESBURY BRIDGE.—Caisson No. 6 in Process of Sinking, Showing Excavator and Shore Chains for Maintaining Vertical Position.

which is entirely wanting in the case of an inclined shoe. This guide is valuable in cases where the soil may differ in density under the shoe, and particularly so if the excavation has been carried too far below the bottom of the shoe. With an inclined shoe and a slip of soil into the dredging well from one side more than another, experience in deep dredging has shown that there is a decidedly greater tendency to a horizontal movement than with a vertical shoe. The former has a flare to direct this sidewise motion in the first place, and nothing but a certain amount of disturbed material above the shoe to resist this tendency."

The above account is from the Engineering News of January 5, 1889, the work having been done under the direction of J. F. Anderson, of the firm of Anderson & Barr. The shells were filled with concrete up to low water and masonry built from low water up to the top of the piers.

Such work may be made water-tight by riveting according to ordinary boiler-maker rules, or if extra thick plates are used this can be exceeded and the rivets spaced some farther apart. The joints may be made with ordinary laps and calked, or a very much better appearance may be obtained by the use of butt joints, and if desirable to avoid calking, then a calking strip may be used to make the joints tight. This is merely a cloth or canvas strip, thoroughly saturated with paint paste, and is laid between the metal surfaces, and the riveting draws the plates upon it and a tight joint will result. The shells will be filled with concrete as soon as the piers are in place and the foundation prepared, so that only a temporary use is required of the strip.

When metal cylinders are used simply as casings for concrete they need not be made water-tight, as they can be dredged out and have the concrete deposited through the water. The metal should never be less than one-quarter inch in thickness, and on first-class

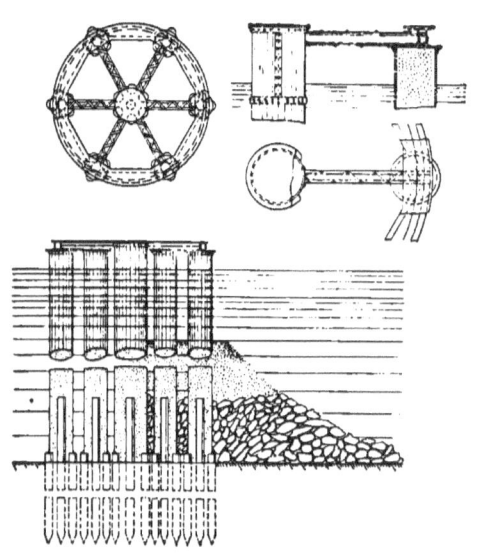

FIG. 58.—GROUP OF CYLINDERS FOR PIVOT PIERS.

work five-sixteenths to one-half inch is preferable. Railroad work of this character is usually constructed of three-eighths inch metal for ordinary depths.

The pivot pier of the bridge over the Little Bras d'Or river in Cape Breton was constructed of seven metal cylinders braced together. The center tube was 4 feet in diameter, while the six outside cylinders were 3 feet in diameter. (Fig. 58). The center pivot, about which the span revolves, rests on the center tube, while the track is supported by the other tubes, but resting directly on rolled beams covered with three-eighths inch plate.

The tubes rest on a clump of piles, cut off at the bed of the stream, with one pile extending up into the center of each tube about six feet, around which the concrete was deposited, thus preventing displacement. Concrete and stone were placed on the outside up to 15 feet, as a protection.

This work was described by Martin Murphy in Trans. Am. Soc. C. E., Vol. 29, who also describes a pier for the Victoria bridge, over Bear river, constructed with two tubes, resting on piles cut off at the bed of the stream, but having four piles inside each tube. (Fig. 59.) Around the outside are timber, concrete and broken stone as a protection. The saw used for cutting off the piles under the water was very much simpler than the one shown in Fig. 35, and is illustrated in Fig. 60.

Cylinder piers on European work are often of very elaborate construction. The bridge on the Aa, at the crossing of the Russian Riga-Orel railway, is supported on elegant cylinder piers, with moulded caps, steel cut-waters, and are braced together with cylinders transversly. (Fig. 61.) This forms a very efficient construction, but so expensive to manufacture that it is usually replaced by bracing of struts and rods, as in Fig. 59, or by a metal diaphragm (Fig. 62), stiffened with angles.

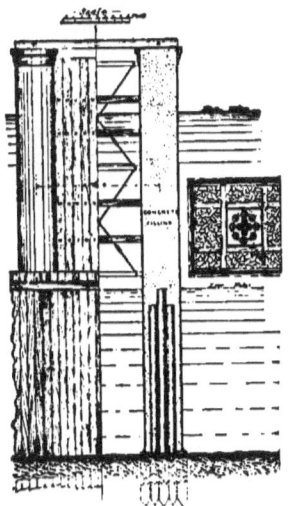

FIG. 59.—PIER OF TWO CYLINDERS, VICTORIA BRIDGE.

Cylinders of water-tight construction and of large diameter may be used as coffer-dams, where they are sunk into impervious strata, or by sealing them with concrete around the bottom where they are placed upon smooth rock bottom. In the construction of light-houses such cylinders have been placed upon clean rock bottom through from 12 feet to 18 feet of water and concrete deposited around the circumference of the base outside and inside to make them water-tight, after which they were pumped out and the foundation laid.

To withstand the pressure of any considerable depth of water the thickness and strength should be calculated and the construction carefully designed. Unless the depth of water exceeds ten feet, or the diameter of tube exceeds six feet, the minimum thickness it is advisable to use, will be sufficient for strength.

This refers only to quiescent pressure, and any shock must be carefully considered and taken account of, by interior bracing if necessary.

The most thorough discussion of the strength of thin, hollow metal cylinders is given in "Elasticitat and Festigkeit," by C. Bach. This considers the cylinder to have sides of a greater thickness than is true with pier shells, and having one radius given, the radius to the other side of the plate is found from the formula, the stress being variable from the inside to the outside of the plate.

For thin cylinders the stress may, without appreciable error, be assumed to be uniform over the cross section of the plate, and the thickness t in inches be found from the formula

$$t = .001\, r\, h$$

where r is the radius of the cylinder in feet and h is the depth of the water

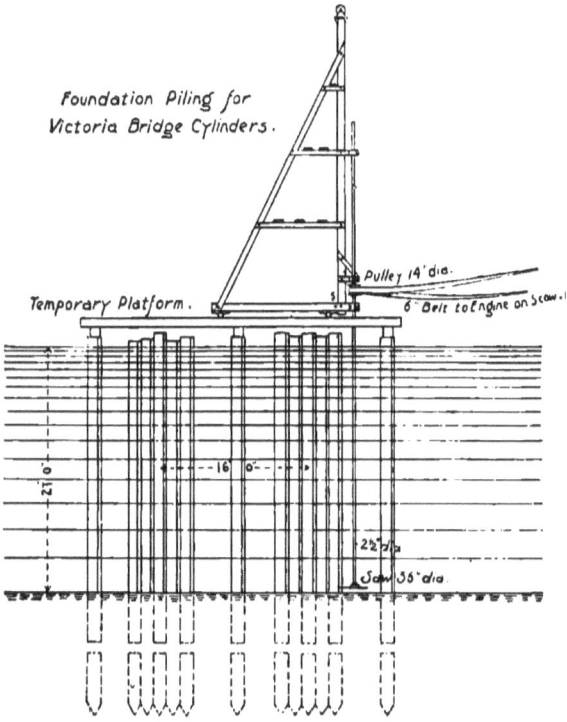

FIG. 60.—CIRCULAR SAW FOR CUTTING OFF PILES UNDER WATER.

to the section in feet, and t in no case to be used less than one-quarter inch in thickness.

This is on the assumption that the metal will stand 5,000 pounds per square inch in compression with safety. For large cylinders, or for rectangular shells, girders and stiffeners or ties and struts must be added to prevent distortion.

The foundations for the great Forth Bridge, which were constructed

FIG. 61.—CYLINDER-PIER BRIDGE, RIGA-OREL R. R., RUSSIA.

under the direction of Sir John Fowler and Sir Benjamin Baker, required the use of various methods to reach solid bearing, as the enormous weight to be carried required the most substantial piers obtainable.

The use of coffer-dams of metal for the Inchgarvie piers is described by Engineering: The site of the two north or shallow piers being wholly submerged at high water, and about half in the case of the northeast and three-fourths in the case of the northwest pier, submerged also at low water, the preliminary work was tidal, and between spring tides no work could be carried on at all at this place. When it is considered how exposed the position was there—the work having to be carried on upon a narrow ledge of rock attacked by wind and waves from all sides—it will be understood that the progress could not be very rapid. The conditions of the contract here required that the rock should be excavated in steps, and that the rubble masonry comprising the foundation of the circular granite piers (Fig. 63) should be bound by an iron belt 60 feet in diameter and 3 feet deep; the highest portion of the rock upon which this belt rested to be 2 feet below low water; the belt, or at any rate a part of it, to be brought down to form a protection for the foundation rubble masonry upon the lower steps.

FIG. 62.—CYLINDER PIERS, WITH DIAPHRAGM.

It was therefore decided to cut a chase 8 feet wide (3 feet to the inside and 5 feet to the outside of the 60 feet circle) out of the rock where it was higher than 2 feet below low water, to make the 60 feet belt of three thicknesses of one-half inch plate and to carry the center plate downward, after it had been cut, in such a manner as to fit as nearly as possible the natural contour of the rock. (Fig. 64A.) A light staging was, therefore, erected above high water, the correct center of the pier placed upon it, and by means of a trammel-rod 30 feet in length, from the end of which a pointed sounding-rod was suspended, a correct reading was taken every 6 inches on

THE COFFER-DAM PROCESS FOR PIERS. 87

the circumference of the 60 feet circle, after a diver had been around to clear out any loose stones lying in the line, or picking off any sharp points projecting. These readings were plotted and the center plates cut to it. In the meantime work had been done upon the chase; and, when nearly cut down to the right level, the belt was put together on the staging exactly above the site of the pier. The plates, projecting downward and forming the shield, were stiffened by I bars vertically over the butts, and where required to be carried down to a considerable depth, as in the case of the northwest pier, they were further stiffened by horizontal circular girders and stayed to the rock by bars of angle iron. The whole belt was now riveted up, and when ready received two coats of red lead paint, and was lowered down to position by means of hydraulic jacks. (Fig. 64B.) The top edge of the 3 feet belt was then leveled all round, and corrected where

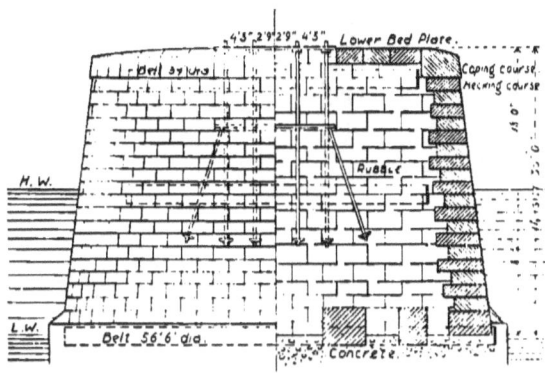

FIG. 63.—CIRCULAR GRANITE PIER AS FOUNDED BY COFFER-DAM. FORTH BRIDGE.

necessary. A heavy angle iron 6 inches by 6 inches by ⅞ inches ran round the inside of the 3 feet belt, and upon this was now set a single tier of temporary caisson, 10 feet in height, and consisting of fourteen segments of about 30 cwt. each in weight. This helped to keep the belt down to the rock, and a number of heavy blocks of stone were placed on the top of the caisson for the same purpose. A sluice door in the lower part was kept open to admit of the tide flowing in and out.

Steps were now taken to make good the joint between the 3 feet belt and the shield and the bed-rock. This was done in the following manner: A number of concrete bags, about 14 inches by 30 inches, and 8 inches to 9 inches thick, were prepared and passed down to a diver, who laid them round the outside of the belt at a distance of about 4 inches. A second row was next laid round the outside of the first row, and tolerably close up,

THE COFFER-DAM PROCESS FOR PIERS.

the space between the two being made up by clay puddle well stamped down. Any split or hole or crevice in the rock was also filled with clay. Upon these two lower rows other bags were now laid crosswise; upon these, two rows lengthwise, and a fourth row crosswise on the top, which was laid close up to the belt. This was done in sections of about 15 feet to 16 feet length all along the shield, but round the outside of the treble belt only two bags deep were laid. On the inside also a single row of clay bags,

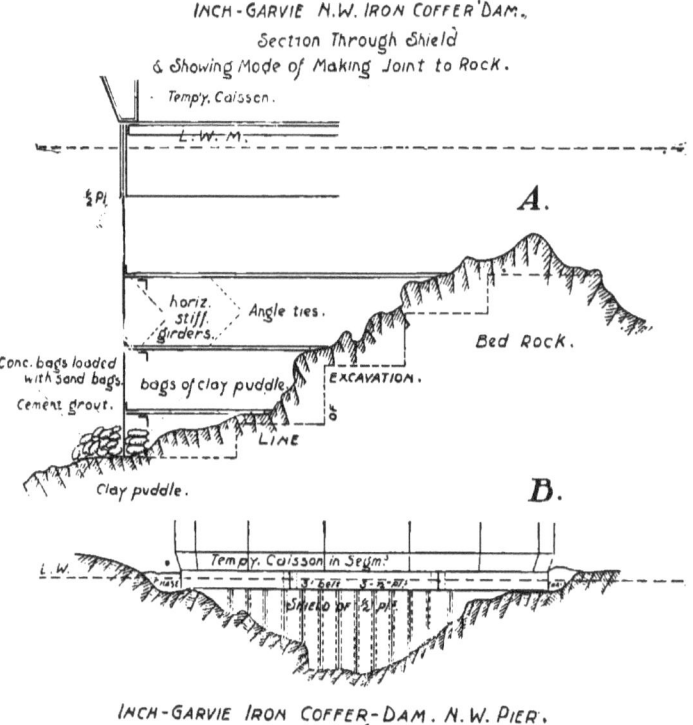

FIG. 64.—FORTH BRIDGE. METAL COFFER-DAM.

backed by a row of concrete bags, and loaded with stones, was laid round the complete circle. Cement grout, without intermixture of sand, was now prepared and passed down to the diver—but only at slack tide, high water or low water—who lifted off one or more of the top bags and poured the grout into the narrow space left, until it overflowed. He then replaced the bag and proceeded to the next division, until all was done. Forty-eight hours were allowed to elapse for the setting of the cement; the sluice valve was then closed and the caisson pumped out gradually. When leaks were discovered the diver descended to examine the outside, and where necessary, cut out some of the grouting and replace it by new.

As it was not considered that this cement joint would be able to stand the full pressure of the tidal rise the coffer-dam was worked as a half tide one, it having to be pumped out every tide as soon as the water had fallen below the top edge of the temporary caisson. In addition to the hydrostatic water pressure, the caisson had to stand the heavy seas thrown against it, whether coming from east or west. Under these circumstances it was often considered advisable not to pump out the coffer-dam, but leave the sluices open and allow the tidal flow free access. Under such conditions it will be easy to see that, during a season of bad weather, much delay could not be avoided, and though the work of excavation had been commenced in the summer of 1883 it was not till the middle of April of the following year that the first rubble masonry could be laid in this pier. In working the excavation no blasting was done within one and a half feet of the iron belt, but the rock was quarried up to within 6 inches and the rubble then built in at once. Any steps cut in the deeper portion were invariably at least twice as broad as they were deep. The deepest point to which the excavation had to be carried in this pier was 8 feet below low water.

The coffer-dam or caisson for the northwest pier, Inchgarvie, was done in the same way precisely as described for the northeast, only that owing to the experience gained by the divers and other men engaged upon the work the progress was much more rapid.

In the northwest pier the depth of the shield was 15 feet below low water, and extended to nearly one-half of the circumference. There was, therefore, in addition to the vertical I bars which covered the butt joints of the shield plates, three horizontal circular girders, carried at a distance of 4 feet 6 inches from each other, and from these a number of horizontal tie bars with cross-bars at the ends were carried radially and level to the rock opposite and pinned to it, and afterward built into the solid rubble masonry. (Fig. 64B.)

This mode of making the joint between the rock and the iron belt was simple and quite effective. Most of the leaks were due to natural crevices in the rock, running from the inside to the outside at a considerable depth. These were circumvented by building small clay dams round, and leading the water by a chute to the pump. Leaks were also caused by the action of heavy waves running up to the temporary caisson at low water with great violence, and shaking the whole fabric.

The whole of the northeast pier was built in a half-tide caisson, as the work was not pressing; but in the case of the northwest pier, so soon as the rubble masonry inside had been brought up to low water level a second tier of temporary caisson was added, and the work could then be carried on at all states of the tide. While tidal work was carried on in these two cofferdams the amount of water which had to be pumped out every tide was 250,000 gallons in the one case and 340,000 in the other. The time occupied was 50 to 55 minutes, but work was, of course, commenced so soon as the higher parts were laid dry. For pumping out smaller quantities of water collected through leaks, pulsometers or small centrifugal pumps were used.

An exterior view of the work is shown in Fig. 65, and while the method

FIG. 65.—FORTH BRIDGE, CIRCULAR GRANITE PIER AND METAL COFFER-DAM.

was successful and worthy of much study, the expense would only be justifiable where the metal would be retained as part of the permanent foundation, which was the case on this work.

In many cases such a shell could be designed of the proper size for the footing course, and after use as a coffer-dam in obtaining the foundation it could be filled with concrete and serve as a base for the pier. Being made in sections vertically, portions projecting above low water could be removed and used on still other piers.

Metal sheet piles are seldom used for any class of work, and need not be discussed at length in this connection. On some harbor work at Cuxhaven Harbor, Germany, hollow metal sheet piles, of elongated elliptical section, were used, and after being driven were filled with concrete.

Whatever the class and form of material it may be decided to use, in securing a foundation by the coffer-dam method, the temporary construction should be so related to the permanent foundation that as much as possible of the material used and labor employed will be of service in the finished structure.

ARTICLE VIII.

THE COFFER-DAM PROCESS FOR PIERS.*

PUMPING AND DREDGING.

THE degree of success which has been attained in the building of a coffer-dam will be evident when the pumping process is begun. After having been pumped out, if the leakage is so small as to require only a small amount of pumping to keep it free from water, it may reasonably be considered a success.

The pumping should not exceed what can be done by a steam siphon, a small pulsometer, or by running a centrifugal pump intermittently. Should leaks develop which cannot readily be contended with, then repairs must be made.

The use of pumps for this class of work on ancient bridges is described by Cresy. The bascule, used by Perronet at the bridge of Orleans (Fig. 66), is one of the most primitive forms. It consists of a see-saw apparatus, at each end of which ten men were placed, and 150 motions were given it in each quarter of an hour. Four cubic feet of water were raised three feet each time, or about 300 gallons per minute. Various other kinds of pumps were used at this bridge, among them the chapelet, which is similar to a modern chain pump, worked by hand. Then the same device was employed, but geared to be operated by horses on a platform. A chapelet operated by a water wheel was also used (Figs. 67 and 68). The large wheel had 124 cogs, while the pinion had 15, which caused the raising of over sixty-six buckets on the chain for each turn of the large wheel. At 180 turns of the wheel per hour, with each bucket lifting 290 cubic inches of water, the capacity was about 250 gallons per minute.

A great bucket wheel was employed by the same engineer at the Neuilly bridge, 16 feet 6 inches in diameter, 4 feet 6 inches wide, with sixteen buckets.

---

* Attention is called to the numerous references in other articles of the pumping plants actually employed on coffer-dams, and especially to the plant used at Topeka, page 78.

Great care should always be given to the selection of a pumping plant of the proper type and proper size, as the statements regarding capacity are often misleading. The outfit should be, if needed, one able to take care of the dredging, if the material is such that it can be pumped.

The pumps used at the present time on very small work are usually square wooden box lift pumps, such as are used on large river barges, and are worked by one or more men lifting on a plunger. These are often replaced by a similar pump of metal (Figs. 69 and 70) with a tube of galvanized metal, and often spiral riveted. The one shown in Fig. 69 has the top and bottom soldered to the tube, while the one in Fig. 70 has screw joints. The cost of a 4-inch pump eight feet long with fixed top and bottom would be about $6, while the screw joints would about double the cost.

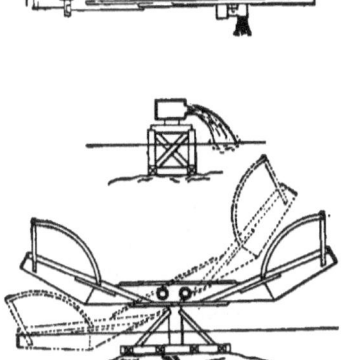

FIG. 66.—OLD BASCULE PUMP.

Such pumps are, however, little used, as the labor becomes excessive where there is any quantity of water to deal with, and diaphragm pumps (Fig. 71) are employed, which work on a rubber diaphragm, in place of a piston and plunger, and throw a large amount of water, besides allowing the passage of sand and gravel without choking the pump. The 2½-inch suction has a capacity of twenty-five gallons per minute, and the 3-inch suction of fifty-eight gallons per minute, the list price of the two sizes being $20 and $26, respectively; the maximum lift of the pump being thirty feet.

Where steam can be obtained steam siphons are often used, the steam being introduced into the main pipe through a nozzle, thus causing a suction, which with a 3-inch discharge Van Duzen jet will deliver 7,200 gallons of water per hour, the height of the pump above water being 11 feet, the point of discharge being 19 feet above the pump, making a total lift of 30 feet. This size will require an 18-horse power boiler and a steam pressure of fifty pounds. The suction pipe is one inch larger than the discharge, while the steam pipe is 1¼ inches in diameter, with a jet opening of about $\frac{13}{8}$ inches.

The list price of a pump of this size (Fig. 72) is $36, the piping being extra. The pump is constructed of gun metal and will last indefinitely. The strainer should always be used and will cost about $4 extra for the 4-inch pipe. The piping should have long bends in place of elbows where a turn is required.

This make of pump is manufactured from ½-inch discharge, with a capacity of 200 gallons per hour, up to 5-inch discharge with a capacity of 12,000 gallons per hour. The smaller sizes are useful for priming centrifugal pumps and for a variety of uses around a contractor's plant.

The Lansdell siphon pump (Fig. 73) has a double suction C C, to which rubber suction pipes are attached. The steam pipe is attached to B, and when the steam is turned on it is blown across A and through D, thus exhausting the air from the chamber A. Water rises through C C by atmospheric pressure to fill the vacuum, and it is forced out through D by the steam, the velocity being proportional to the steam pressure. The steam

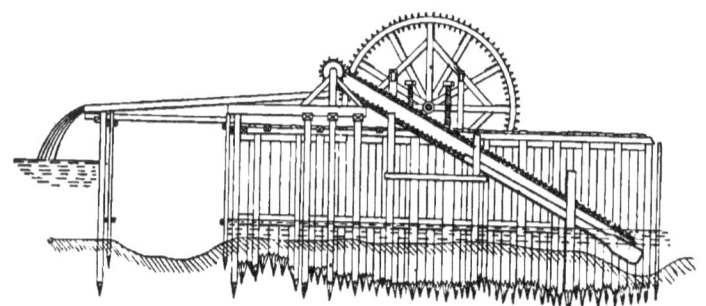

FIG. 67.—OLD CHAPELET, SIDE ELEVATION.

supply should be as close to the pump as possible, to prevent condensation, and the turns in the pipe should be easy bends, as stated regarding the Van Duzen jet. When the height exceeds fourteen feet, to which the water is to be pumped, the suction pipes must be long enough to allow the center of the pump to be placed fourteen feet above the water. With a 3-inch discharge, a 1½-inch steam pipe is required and a 12-horse power boiler. With a 6-inch discharge a 2½-inch steam pipe is required and a 50-horse power boiler.

The rated capacity of the 3-inch is 450 gallons per minute, of the 6-inch 1,800 gallons. But this would likely not be realized in practice.

The vacuum pump which has reached the most general adoption is the Pulsometer, and is in many ways better adapted to light service than a centrifugal pump of small size. There are no bearings to keep up, no belts to keep tight, and no trouble in preparing a foundation, as the pump is suspended

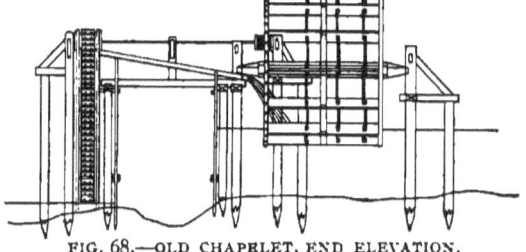

FIG. 68.—OLD CHAPELET, END ELEVATION.

by the hook shown in Fig. 74. The pump is operated by admitting the steam through the pipe at the extreme top (Fig. 75), the pump having been previously primed by filling the middle chamber with water. The air valves are closed and the steam passes into the right hand chamber A

THE COFFER-DAM PROCESS FOR PIERS.

FIG. 69.—HAND PUMP. SOLDERED JOINTS.

FIG. 70.—HAND PUMP. SCREW JOINTS.

clearing it of water by forcing it into the discharge chamber shown in dotted lines. The steam then condenses at once and the ball C changes its seat, closing the right hand and opening the left hand chamber to the steam. The vacuum, formed by the steam condensing in the right hand chamber A, allows it to fill with water by atmospheric pressure through the suction pipe at the extreme bottom and through the chamber D, it being retained by the valves E E. The steam then enters the left hand chamber A and the operation is repeated. The chamber J is a vacuum chamber.

In starting the pump the steam is turned on for three or four seconds, then shut off for four or five seconds, alternating these movements until the pump is started. The steam is then turned on about half or three-quarters of a revolution, the two side air valves opened about half a turn, and then the middle air valve opened slowly until a regular stroke is obtained.

The capacity of the 3-inch discharge, with a ½-inch steam pipe and operated by a 9-horse power boiler, is 180 gallons per minute when the lift is as much as twenty-five feet; and for the 6-inch discharge, with a 1½-inch steam pipe and operated by a 35-horse power boiler, 1,000 gallons for the same lift.

The pulsometer is remarkably smooth in operation, and except for the slight click of the ball and the discharge of water in a steady stream, one would scarcely know it was pumping. Where a good-sized hoisting engine boiler is in use on foundation work, it can be used to supply the steam for pumping. The work illustrated in Fig. 4 was easily kept free of water by a small pulsometer.

FIG. 71.—DIAPHRAGM PUMP.

while its use has been cited in a number of cases where the cofferdam was pumped out by a centrifugal pump, and then the leakage kept under control by a medium sized pulsometer, which required but little attention. The pump should be provided with a strainer at the bottom of the suction pipe, all the connections must be air tight, no sharp bends should be made in the pipe, and with dry steam successful working will result. Another pump of similar construction is the Maslin Automatic Vacuum Pump, which differs from it in important details. What has been said regarding the pulsometer will apply as well to the Maslin pump.

All the foregoing devices are for use where the amount of water to be handled in a given time is of limited amount, but where large quantities are to be pumped out of cofferdams in short periods of time, resource must be had to centrifugal pumps, which have reached a high state of perfection. Where the water is to be lifted ten feet an ordinary reciprocating pump would exhibit an efficiency of only 30 per cent, while a centrifugal pump would have an efficiency of 64 per cent. For a lift of seventeen feet the reciprocating type would have an efficiency of 50 per cent, while the centrifugal would reach its maximum of 69 per cent efficiency, dropping to only 50 per cent for a lift of fifty feet, while the other type would increase to 75 per cent. From this it will be seen that the centrifugal pump is essentially a low lift machine.

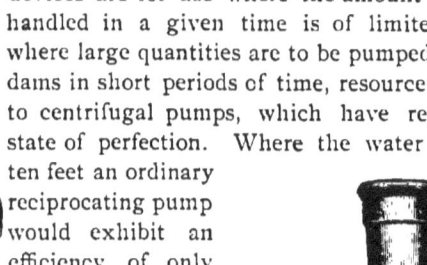

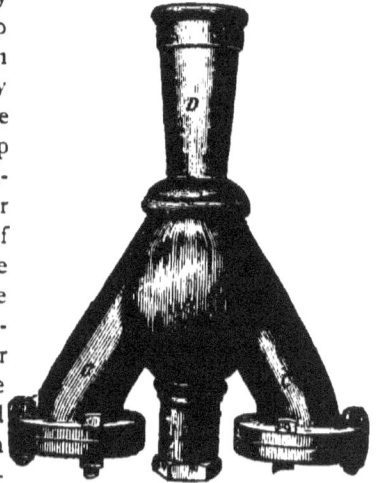

FIG. 72.—VAN DUZEN JET PUMP.

FIG. 73. LANSDELL'S SYPHON PUMP.

Actual tests of pumps show that the maximum results are very seldom realized, a 9-inch discharge of one make showing an increase from 46.52 per cent for a 12.25 feet lift, to 57.57 per cent for a 13.08 lift; while another make of 10-inch discharge, shows a decrease from 64.5 per cent for a 12.33 lift, to 55.72 per cent. for a 13 feet lift. The greatest efficiency at hand is

THE COFFER-DAM PROCESS FOR PIERS. 97

shown by a German pump with a 9¼-inch discharge, a 10.3-inch suction, and a 20.5-inch disk, running at 500 revolutions. The lift was 16.46 feet and the efficiency 73.1 per cent!

That such results are not realized on actual work is readily understood when it is considered what little care is used to properly place and operate such a plant, how little attention is paid to having a proper boiler and engine, and what lack of care there often is to keep the plant in good repair.

An ideal outfit for operating by steam is shown in Fig. 76, where the engine is directly connected to a Heald & Sisco pump. All the trouble and vexation

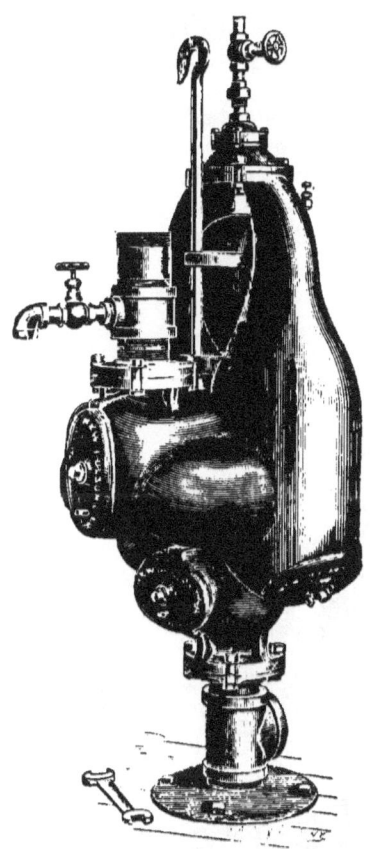

FIG. 74.—PULSOMETER STEAM PUMP.

from the use of a belt being done away with, and no loss of power through slipping of belts. The machine can be placed on the barge which carries the boiler, the suction pipe being run horizontally across as in Fig. 56, while a short discharge pipe discharges directly into the river. Where electric power plants are available a still better arrangement will be to have an electric motor directly connected to the pump, and all the trouble incident to the use of a boiler on the work will be avoided.

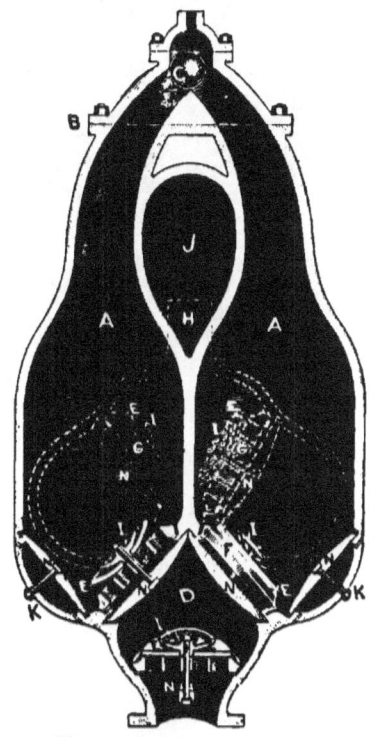

FIG. 75.—SECTION OF PULSOMETER.

Electric power can also be used for hoisting and for pile driving. Examples of the use of motors on hoisting machinery will be given in a later article.

The suction should always be fitted with a section of smooth-bore rubber hose (Fig. 77A) to give it flexibility, a length of about eight feet being usually sufficient. The best hose is made with a spiral metal core which adds to its strength and durability.

The suction pipe is ordinarily made of sections of wrought iron pipe,

FIG. 76.—CENTRIFUGAL PUMP, DIRECTLY CONNECTED TO ENGINE.

with screw connections, but as this is troublesome to change sections, it will be found advantageous to use the spiral riveted pipe with flange couplings (Fig. 77B), and to have extra sections from two to six feet long, with several sections of each shorter length, so the length of the suction pipe can be readily changed to suit the depth of the excavation. The flanges must be provided with rubber gaskets to keep the pipe air tight.

The strainer (Fig. 77C) is used to prevent large stones, sticks or obstructions from entering and clogging ordinary pumps, and usually comprises a foot valve to retain a pipe full of water and make the priming easy. The strainer or end of the suction pipe is usually placed in the lowest point, and sometimes a box or sump is provided, as a well into which the water is drained from the other and higher portions of the work. A small set of falls should be attached to the foot to raise the pipe and clean out the strainer when necessary.

The centrifugal pump itself must be in first-class repair to do economical work, and should be a large enough size so that it need not be run beyond its economical capacity. The style of pump to use will depend upon the work to be done, but for coffer-dam work a vertical pump could not be used easily and need not be considered. Where practically clean water is to be pumped an ordinary style of pump should be used, but where much mud or sand will be drawn up a sand pump is best; and where a large part of the excavation is to be done with the pump, as at Topeka, a dredging pump will be the proper type.

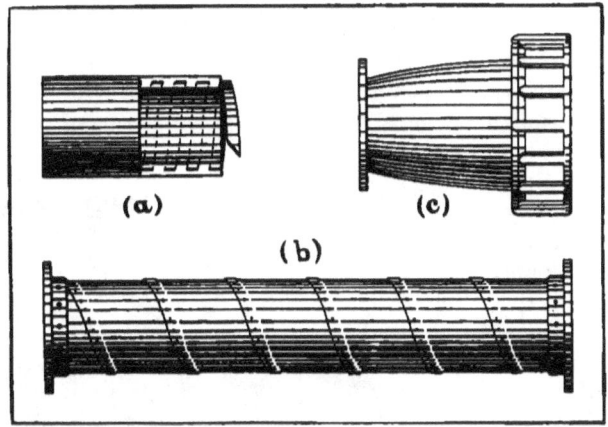

FIG. 77.—SUCTION DETAIL FOR PUMP.

The pumping required on the Chattanooga work, 5,000 gallons per minute to a height of about fifteen feet, would have been done most economically by a 15-inch pump, with a 40-horse power engine and a 50-horse power boiler. But a pump of this size would not find ready use in a contractor's work, and for this reason two 8-inch pumps would have been the better outfit to purchase, unless the work was very extensive; and each pump should be provided with a 25 or 30-horse power engine, so as to run the pumps somewhat beyond the economical capacity, which could readily be done with a direct connected engine, where there would be no belt to slip.

The work required on the Forth bridge coffer-dams could also be done by the 15-inch pump above described, the lift being about 3 feet at the start and reaching 18 feet as the dam was cleared, the 340,000 gallons being pumped out in about one hour.

Centrifugal pumps are rarely required for a lift of over 20 feet on this class of work, which is only slightly beyond the economical lift, and the height should never exceed 30 feet, which would require for the 15-inch pump an engine of 75-horse power.

FIG. 78.—CENTRIFUGAL PUMP, DOUBLE SUCTION.

The pump may be located on the coffer-dam, but in case of high water during the progress of the work the outfit may be damaged and it is best to place the pump on a boat, as in Fig. 56, with a section of horizontal suction pipe across to the work, which should be as short as possible.

The ordinary type of pump (Fig. 76) may be fitted with a primer, consisting of a small hand force pump attached to one side of the pump, for filling the pump and suction pipe. A more simple way is to provide a barrel above the pump, which can be kept full by using a small steam jet, and by means of a pipe with valve from the bottom of the barrel to the top of pump, the contents can be emptied into the pump to prime it. Priming may also be easily accomplished by inserting a hose into the discharge pipe and filling the pump directly with a steam jet.

FIG. 79.—DREDGING PUMP.

Double suction pumps (Fig. 78) allow the water to enter on each side of the piston, and thus a perfect balance is secured, which does away with all end thrust on the bearings. This pump is most easily primed by using an ejector, or a flap valve such as is shown on the discharge pipe of the dredging pump (Fig. 79) and which serves to retain the water in the pump. Where a long discharge pipe is to be used, a quick closing gate valve may be introduced into the pipe near the pump.

Where the material to be dredged out at the foundation site is mud or sand or partly gravel, it can be removed during the process of pumping by using a dredging pump. In case there were 700 yards of material to be removed and an 8-inch pump was provided, it would not be advisable to count on more than 10 per cent. of solid matter being discharged by the pump, as the suction could not be kept working close up to the sand or mud. By using a 30-horse power engine, a discharge of 2,000 gallons per minute would be reached, or with 10 per cent of loose solid matter, the excavation would be made in less than two working days.

FIG. 80.—DREDGING PUMP PISTON.

The piston of a dredging pump (Fig. 80) is provided with large openings to receive the material, and the one illustrated is provided with side plates so that all wear is taken off the pump casing.

One of the most remarkable pieces of work done with this class of pumps was the use of Edwards' Cataract pumps in dredging the ship channel in New York harbor. This is described in the Trans. Am. Soc. C. E., Vol. 25. The work was done by three dredges, which were much the same as small sea-going vessels, the largest being the Reliance, 157 feet long, and carrying 650 cubic yards of dredged material. Two separate pumps were provided, each with 18-inch suction pipes, reaching from the sides of the vessel and parallel to it down to the bottom to be dredged, being supported by suitable hoisting tackle. These boats were kept under headway toward the dumping ground while the dredging was in progress. The average load during about a month's working of the Reliance was 585 cubic yards and the average time of loading about 48 minutes, while the average number of loads per day was 6.73.

These dredges removed the enormous quantity of 4,299,858 cubic yards of material at an average price of 24.48 cents per yard, the lowest price being about 17 cents, the average price paid for other forms of dredging being

40.53 cents. On foundation work the amounts to be removed would be small and the cost for this reason much higher, yet owing to the smaller cost of the plant that would be required, the cost need not be greatly in excess of the above. It is usual, however, as the amount to be dredged will cost such a small proportion of the total cost of the substructure, to figure from $1 to $2 per yard for excavation in ordinary coffer-dams.

Reference has already been made to hand dredging and a very cheap and effective scraper was illustrated in Fig. 8. Where dredging is to be done in tubes, wells or puddle chambers, it can be done by a clam-shell dredge or grapple such as was shown in Fig. 57, in use on the Hawkesbury foundations.

The Lancaster dredge (Fig. 81) is a well known form of this type of

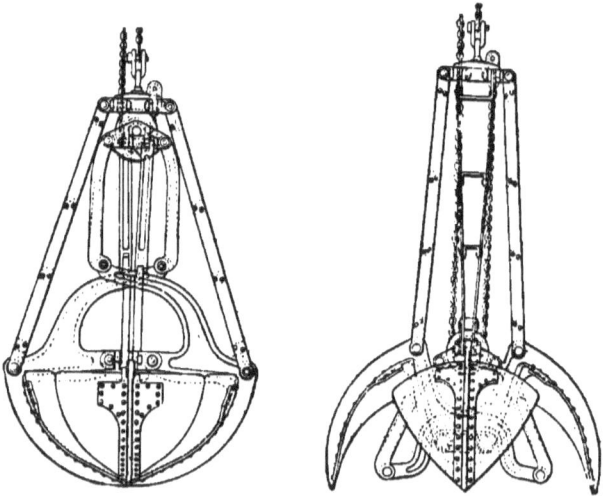

FIG. 81.—LANCASTER GRAPPLE.

machine, and can be operated from an ordinary derrick which is served by a double-drum hoisting engine. This dredge will work best of course where there is some depth of soft material to be removed. While a large dredge would generally be hired by a contractor, these buckets can be owned by him and the work carried on cheaply and conveniently.

Sand diggers such as were mentioned in Article II can often be hired where other means are not at hand, or they can be rigged up very cheaply if necessary. A very simple one (Fig. 82) can be built on an ordinary barge, the engine being an ordinary one with a vertical boiler, while the buckets are mounted in a very simple manner and operated through a well in the center of the boat. Such a dredge will dig about 100 yards of sand per day, with only two men to attend it, and will use less than one-half ton of cheap coal, the total cost per yard thus running below five cents. Large elevator dredges of this type are very elaborate affairs, and as they are in wide use they can often be hired for making excavations.

The best known form of dredge, perhaps, is the dipper dredge. The Osgood machines (Figs. 83 and 84) in use on the New York State canals are among the best machines of this kind in use. Such dredges are more simple in construction than elevator machines, and are consequently easier and cheaper to keep in repair. The hull is 70 x 17 x 6 feet with two 6-feet pontoons which are removed when going through locks. The engines consist of a double drum main engine with 8x10 inch cylinders, a swinging engine with 6x8 inch cylinders, and a crowding engine, 5x6 inch cylinders, which are all used in operating the digger of 1¼ yards capacity on a steel boom 45 feet in length.

FIG. 82.—SAND DIGGER.

The crowding engine is used to control the dipper and enables it to make a practically level bottom at one cut, and also thrusts the dipper far enough beyond the boom to allow it to dump fifty-two feet from the center. This dredge, which cost complete $10,000, is operated by a crew of only four men and consumes but one ton of coal per day of twelve hours, the average excavation during four months' work being 549 cubic yards per day. The machine has sufficient power to dig hardpan, boulders, and very soft shale rock.

A dredge of this make, of 3½ yards capacity, working in mud and sand, has dug material at the very low actual cost of .99 of one cent! This of course was an exceptional case, and the cost will rarely fall below five

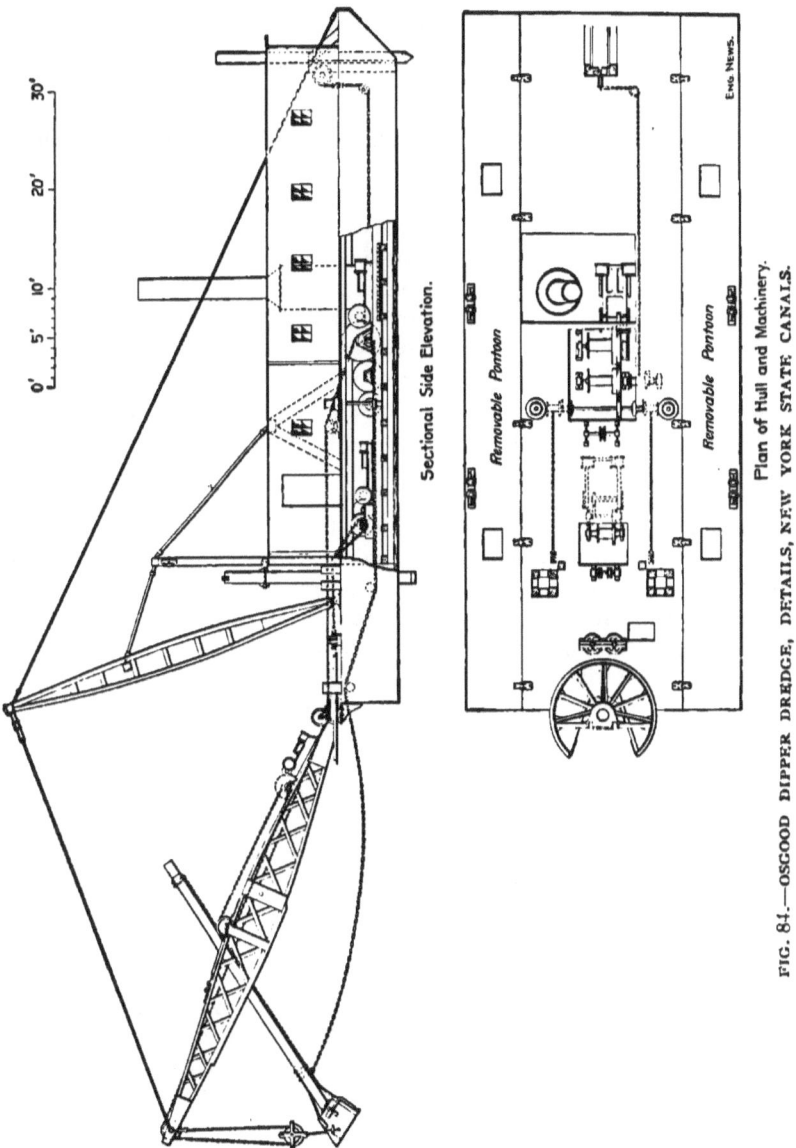

FIG. 84.—OSGOOD DIPPER DREDGE, DETAILS, NEW YORK STATE CANALS.

cents per yard on easy work at a depth not exceeding ten feet, and in such small amounts as would have to be dredged on coffer-dam work and in about twenty feet of water the actual cost would likely reach fifteen cents per yard. In case the dredge should be hired to do the work, a charge of from twenty to thirty cents per yard would not be excessive depending of course on the class of material and the amount.

ARTICLE IX.

THE COFFER-DAM PROCESS FOR PIERS.

THE FOUNDATION.

THE coffer-dam is only the means of reaching a desired end, and this must be borne in mind and the construction made as simply as possible to obtain a first-class foundation.

When the coffer-dam is completed and pumped out work can then proceed if the pumps are able to control the water easily. The character of the foundation having been previously decided upon, after a careful examination of the site, it is assumed that the temporary work has been executed in a manner which is properly related to the permanent structure.

The different kinds of bottom likely to be encountered are: First, light sand and gravel or mud of unknown depth; second, similar material overlying either cemented gravel, clay, hardpan or rock; third, a clean rock bottom, which is approximately smooth and level; fourth, a sloping rock bottom, which is either smooth or rough, and fifth, a rough and irregular rock bottom.

Should the bottom be of the first kind—light sand and gravel or mud of unknown depth—the soft upper layer may have been removed by a dredge previous to the building of the dam, or it may be removed by a dredge or grapple from within the inclosed area, and without the necessity of keeping the dam pumped out, or pumping may be kept up with a dredging pump and the light material removed in this way, after which the heavier material may be removed as deep as necessary by hand shoveling and a dirt box, as shown in Fig. 56. In such a bottom the foundation is usually made by driving piles from two to four feet centers, this distance being regulated by the bearingpower, as determined from Wellington's formula in Article IV, and building upon the tops of the piles, after they have been cut off to a level below low water, a grillage of timber. The space between the piles should be filled with broken stone or concrete, and the grillage placed entirely below low water, the coffer-dam being kept pumped out to allow this work to be done, and also during the laying of the footing courses of the masonry which are below the water.

When the soft bottom overlays good clay, hardpan or rock, as in the second case, and the depth exceeds 20 or 25 feet below the water surface, piles may be driven to the harder substratum and act as bearing piles. But when the depth is in the region of 20 feet or less, it is best to excavate and place the foundation masonry directly upon the solid bottom. The foundation will be of the character described for some of the following cases:

The third class is similar to the foundation at Chattanooga after the gravel was removed. The fissures in the rock are filled up or closed with cement and concrete, and a leveling course of concrete put down on which to found the pier. (Fig. 49).

Bottoms of the fourth class should have all the loose and decomposed rock removed and steps cut out by blasting and wedging, to give a secure hold for the foundation, but if it is simply rough and irregular a leveling course of concrete will be all that is required on which to start the pier. Bottoms of clay and hardpan will require a very similar treatment, except that the leveling course of concrete must be made of sufficient thickness to properly distribute the pressure, which will seldom be less than three feet and can often be increased with advantage to six or eight feet. An example of the stepping of rock bottom was given in the account of the Forth Bridge piers in Article VII and was shown by the dotted lines in Fig. 64.

Where there is a current caused by leakage through the sides of a coffer-dam, or from the bottom, or if the water within the dam is agitated by the pumping, it will be best, after the bottom is clean and properly prepared, to allow the water to run in and then deposit the concrete through the still water. This has been successfully accomplished through 25 or 30 feet of water, and while some engineers recommend allowing the concrete to set from one to three hours before depositing, to prevent the cement from washing out of the concrete, this is not necessary nor advisable if the proper care is exercised and the proper apparatus used. The concrete should be made from one-third to one-half richer than would be used for similar open air work, as there will be some loss of strength.

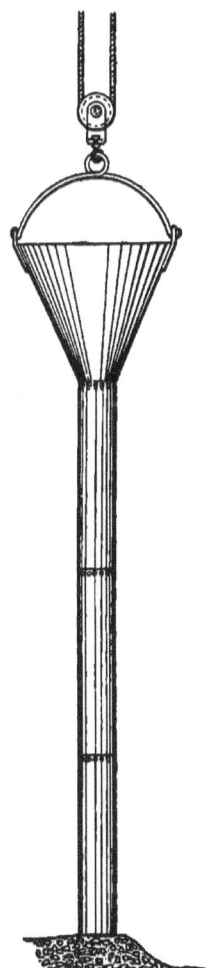

FIG. 85.—METAL TUBE FOR CONCRETING.

The simplest method is to deposit the concrete in paper sacks by sliding them down a smooth wooden or iron chute, or by loading them into a box or skip and dumping them out after the box reaches the bottom. The sacks should be of tough paper, similar to flour sacks, and when they reach the bottom they may be broken by a pike pole and the concrete allowed to run

108 THE COFFER-DAM PROCESS FOR PIERS.

together. Thin cloth sacks are sometimes used and they become fairly well cemented together by the mortar which oozes through.

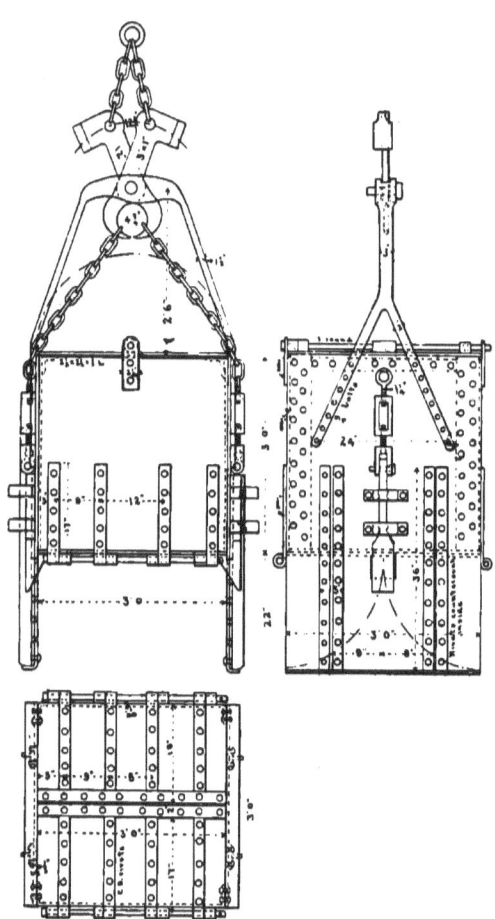

FIG. 86.—METAL BUCKET FOR CONCRETING.

Where the amount of concrete is considerable it will be best to use a tube or bottom dumping box. For placing concrete under water on the Bouci-

THE COFFER-DAM PROCESS FOR PIERS. 109

cault bridge over the river Saone in France a wooden tube 16 inches square was used. This is described in the Engineering News of May 18, 1893. The tube was carried transversely across the caisson on a traveling crane which ran lengthwise of the caisson on tracks on the sides, thus allowing the tube to be moved in any desired direction. The tube was built in sections which could be easily removed, was provided with a hopper at the top into which the concrete was dumped, and a drop door at the bottom to let

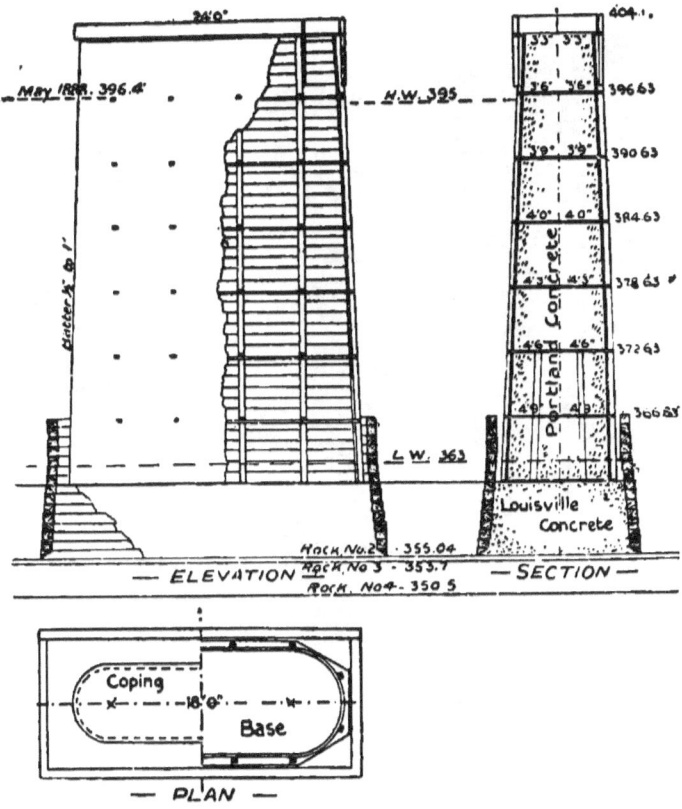

FIG. 87.—CONCRETE PIERS. RED RIVER BRIDGE.

out the concrete. The tube was filled as it was lowered down into the water, and opened when within 16 inches of the bottom. As concrete was dumped in above, the tube was moved about and a 16-inch layer of concrete deposited. When one layer was complete, another of the same thickness was deposited. This method of using 16-inch layers was said to have obviated laitance or the exuding of the gelatinous fluid which prevents uniform setting. The concrete was deposited about the heads of the piles and no grillage used. The thickness of the concrete, which was deposited at the rate of from 90 to 100 yards per day, was 9.84 feet, and was allowed to set fourteen days before the pier was begun.

A metal tube may be used, such as was employed on the Harvard bridge at Boston by W. H. Ward. This tube (Fig. 85) was not provided with a bottom and the first filling of the tube was consequently done after the tube was lowered and the concrete became somewhat washed. This may easily be prevented by using concrete in paper sacks to fill the tube the first time. The tube was suspended from a derrick and was moved about so as to keep the concrete level and deposit it in layers. This account is taken from Vol. 31 of the Engineering News, from which is taken the following description of a metal bucket used by W. D. Taylor on the Coosa river.

FIG. 88.—CONCRETE FORMS. RED RIVER BRIDGE.

This bucket (Fig. 86) was of riveted construction and held one yard of concrete. The maximum depth of water was 26 feet, at which depth the bucket and its load became so lightened that the bucket tripped as soon as the flanges touched the bottom. Similar boxes are often constructed of wood, or they are often made "V"-shaped, one side being arranged to open and dump the load.

For concrete work of this character natural cement is often used, but on all important work Portland cement should be employed. The proportions range from 1 of cement, 2 of sand and 4 of broken stone, to 1 of cement, 3 of sand and 6 of broken stone.

On such a base either a masonry or monolithic concrete pier may be constructed. The pier at Little Rock (Fig. 52) was of this construction and of the composition given in Article VI. A similar piece of work was constructed on the Red River bridge on the St. Louis & San Francisco Railway and is described in the Engineering News of June 2, 1888, by C. D. Purdon, assistant engineer in charge of the work, under James Dun, Chief Engineer. The cribs were filled with Louisville cement concrete up to within two feet of low water, on which was built the pier. (Figs. 87 and 88.)

"After the crib had been filled with concrete and the surface leveled off, the center lines of the pier were located and a frame of 2"x8" plank the shape of the pier, and four inches larger to allow for lagging, was placed in exact position and held by pieces spiked to the crib. On this frame upright posts 6"x6" and 5' 10" high, with a batter of one-half inch per three feet were set in the position shown on the drawing, then the feet spiked to the frame and another frame similar to the first, but six inches narrower placed on them. This again was brought to exact position and braced to the crib

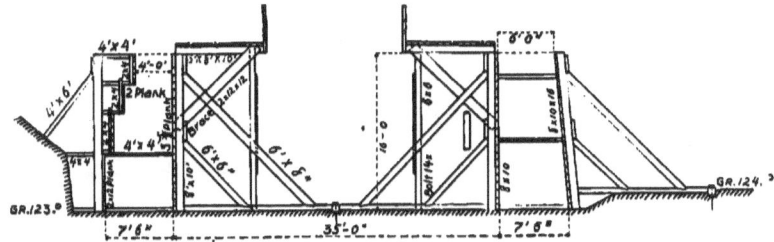

FIG. 89.—CONCRETE FORMS. ILLINOIS AND MICHIGAN CANAL.

and the frame completed by putting lagging of 2-inch plank inside the posts and spiking to them. This lagging was horizontal in the body of the pier and vertical (2"x4") at the ends, beveled pieces being introduced in the ends at intervals to make up the difference of the upper and lower circles. Next 2"x6" planks were placed across on the top of the posts, running clear through the pier, to act as braces. In the rest of the frames these braces were allowed to extend about six feet on each side and the frame braced by spiking plank to them and to the vertical posts. After a section of frames was completed a bed of cement mortar about two inches thick was spread all over the concrete in the crib. On this the rough stone, in such pieces as one man could easily handle, was placed so that no two pieces would be closer than two inches, nor any piece within two inches of the frame, the stone being thoroughly wet before laying.

"Next, on this course of stone another bed of mortar was placed, sufficient to fill all the spaces between the stones and remain about two inches thick above them. It was then well rammed with rammers made by inserting a handle in a section of a pile, except at the edges, where a rammer made of a 2-inch plank cut in the shape of a spade was used, to insure a perfect skin of cement without any breaks. After this had been well rammed, another layer of stone was placed and covered with mortar as before, and so on.

"The coping, which was made similar to the body of the pier, was finished by about 1½ inches of cement mixed with sand one to one, fluid

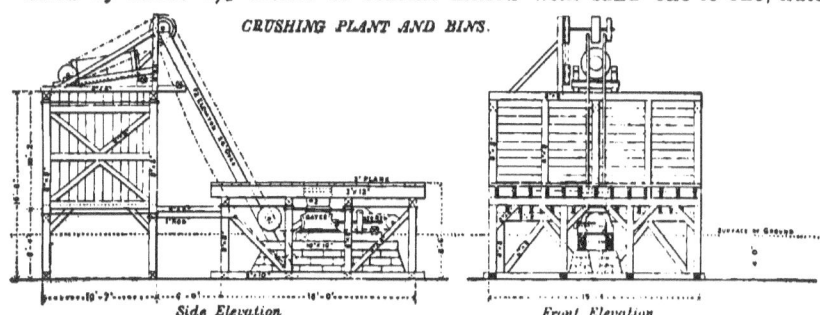

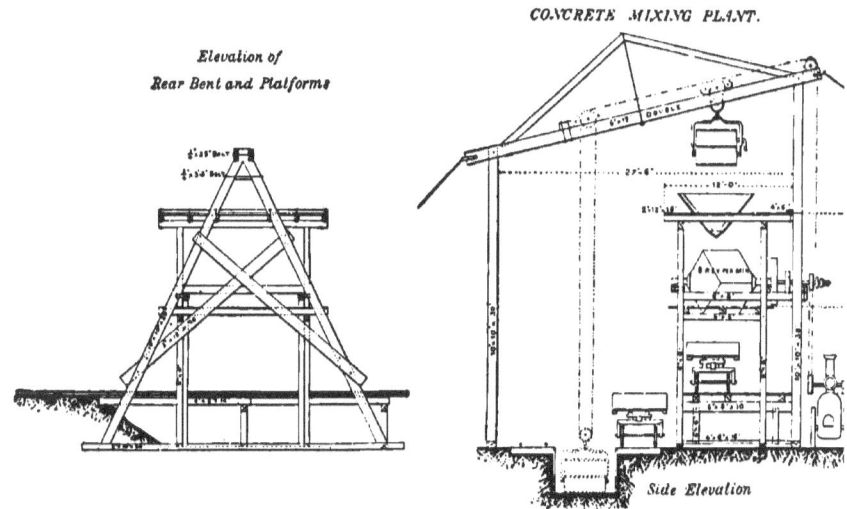

FIG. 90.—STONE CRUSHER AND CONCRETE MIXER. ILLINOIS AND MICHIGAN CANAL

enough to be struck off by a straight edge, the top of the frame being dressed and leveled for that purpose.

"After the pier had been completed the frames were removed and the braces running through the piers cut off by a chisel inside the concrete. Then, to make a smooth surface, the pier was thoroughly wet and plastered with a mixture of one part sand to one part cement, after all the rough or loose portions had been scraped off. This was mainly done for appearance."

The mortar for the body of the pier was made of one part Alsen's German Portland cement and four parts of sand. There was used about 1⅓ barrels of cement to a cubic yard of completed pier. In mixing the mortar eleven ordinary pails full of water were used to one barrel of cement, which caused the water to just appear on the surface when the tamping was done.

The lock walls on the Illinois and Mississippi canal have been constructed of monolithic concrete under Captain W. L. Marshall, Corps of Engineers. The work was executed under L. L. Wheeler, engineer in charge, from whose account in the report of the Chief of Engineers for 1894, the following is taken:

"The rules adopted for the work were adhered to and are worthy of careful study.

"I. The forms or molds of the walls will be divided by vertical partitions perpendicular to the longest axis of the mass, and the walls be constructed by filling alternate sections.

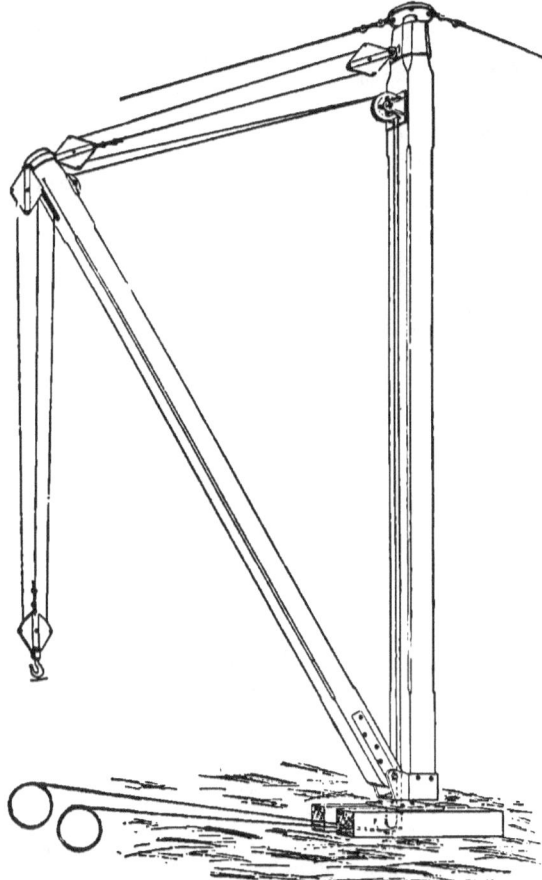

FIG. 91.—AMERICAN HOIST AND DERRICK CO. DOUBLE DRUM GUY DERRICK.

"II. The sections will be filled in horizontal layers, well rammed, each layer to be deposited before the 'initial set' of the previously deposited layer. When the work of filling a section is begun it must proceed without intermission to completion, working night and day if necessary.

"III. The facing and backing must go on simultaneously in the same horizontal layers, using the same cement in the facing as in the backing, with no defined lines of demarcation between the facing which contains no stone and the concrete backing.

"IV. When the top surface of the coping is reached it will be finished after ramming by cutting off the excess by a straight edge, and rubbed smooth and hard by a float. No plastering or wet finishing coat will be allowed.

FIG. 92.—SINGLE DRUM HORSE POWER. CONTRCATORS PLANT MFG. CO.

"V. The facing of the walls will not be finished by plastering or washing with cement after the forms are removed, nor dressed in any manner beyond chiseling away rough ridges should the plank forming not be smooth.

"VI. The concrete shall be mixed with all the water it will take, without water showing after ramming, or without 'quaking' upon ramming.

"VII. At such intervals as may be necessary vertical wells, at least one foot square will be formed along the middle of the masses of concrete, reaching to near the bottom thereof. The masses of concrete after forming will be kept sheltered from the sun, the outer surfaces kept moist and the wells kept filled with water until well set, or about three weeks. The walls will then be filled with concrete.

"VIII. In preparing the cement for mixing with other ingredients of concrete, from five to ten barrels will be kept thoroughly mixed dry, to guard against chance barrels of defective cement, and the necessary quantity of cement will be taken for each batch from this mixture.

FIG. 93.—DOUBLE DRUM HOIST ENGINE. LIDGERWOOD MFG. CO.

"IX. Two cements of different qualities shall not be used in the same section, but as far as practicable each mass shall be homogeneous throughout, but a slight excess of cement in the facing to reduce its capacity to absorb water."

THE COFFER-DAM PROCESS FOR PIERS. 115

The rate at which the concrete was deposited in the work was determined by the rate of ramming, and but one yard every five minutes was deposited. The forms (Fig. 89) were lined with dressed pine plank 4 by 8 inches on the face, of uniform thickness, and with 2-inch rough plank on the back.

Rough plank is sometimes used on such work and lined with oiled paper, or ordinary dressed plank may be used and coated with soft soap. In most sections of the country crushed broken stone can be obtained, but owing to the magnitude of this work a crusher was built (Fig. 90) and was found to work very satisfactorily. The concrete mixer shown in Fig. 90 was operated by a 15-horse power portable engine. The proportions finally adopted for the concrete were one of cement, three and one-third of gravel, and four

FIG. 94.—LIDGERWOOD ELECTRIC HOIST.

of broken stone, while the facing and coping were composed of one part cement and two parts of clean river sand.

That the sand for concrete be clean and sharp is very essential, and any loam or dirt must be washed out. Equally essential is good, clean, sharp, broken stone without dust or dirt. The cement used on the above work was a German Portland, but several of the American Portlands are first-class and will give as good results as the imported.

Where good, fresh cement is being supplied, a few tests to a carload will be sufficient, and for cements of the grade of Atlas or Empire, the guarantee of the manufacturer, supplemented by a few tests, should be sufficient. But for cements which have been shipped by water tests should be made from every five or ten barrels.

The Atlas Cement Company recommend, for concrete laid in open air on moist ground where great weight must be carried, one of cement, two of clean sharp sand, and four of 2-inch broken stone; this sand and cement to be thoroughly mixed dry, then just enough water added to thoroughly moisten, and the mass turned over at least twice, when the stone is to be

FIG. 95.—LIDGERWOOD CABLEWAY CARRIAGE AND SKIP.

added in a thoroughly wet condition. This must then be put at once into the molds and well rammed.

Where a solid bottom is to be built upon, the proportions of one of cement, three of sand and six of broken stone are recommended. For ordinary construction one of cement, four of sand and eight of broken stone, while to obtain a concrete as strong as ordinary natural cement concrete, one of cement, five of sand and ten of broken stone can be used.

The average cost of such concretes, including labor, tools, timber forms and a fair profit to the contractor would be for the first $8 per yard, for the second $7, for the third $6, and for the fourth $5.

Where the leveling course of concrete has been put in and the pier is to be of stone, the footing course should be of carefully selected material. They should be large stones with good beds, and should be as thick or preferably thicker than the courses above. Where the bearing pressure does not exceed two tons per square foot, the footing courses may be stepped by allowing each course to project about one and one-third times its thickness, depending of course on the quality of the stone.

The usual way of handling the material for foundations and piers, is to boat it to the site, where it is placed by a stiff-leg derrick, or if guys can be used, by a derrick with wire rope guys. The fittings for such derricks can be obtained from a number of firms, an American Hoist and Derrick Company outfit being shown in Fig. 91. This is rigged to be operated by a double drum hoist, which can be one operated by horse power (Fig. 92) if the piers are near the bank and if steam power is not available. The usual form, however, is a double drum steam hoist like the Lidgerwood machine shown in Fig. 93. Where electric power is available an electric hoist (Fig. 94) should be used, as it will be found much more convenient.

Works of any magnitude should, however, be fitted from the beginning with a cableway, which will avoid the necessity of boating the materials, erecting of large derricks, and facilitate in every way the prosecution of the work, besides often making a balance on the right side of the ledger. The Lidgerwood cableway on Dam No. 11 of the Great Kanawah river, a tower of which can be seen in Fig. 9, had a span of 1,505.5 feet and carried a net load of four tons on a main cable 2½ inches in diameter. The stone quarry was located on one bank and the stone was taken directly to the stone yard and to the work in the river. A seam of coal in the quarry also supplied fuel for the dredges and pumps, the coal being handled by the cableway, as was also the material from the railroad siding on the opposite bank.

The details of these cableways have been developed and perfected to a wonderful extent, as a result of their use on the Chicago Drainage channel. The engine for operating one of these with a capacity of eight tons has double 10x12 cylinders, the cranks being set at an angle of 90 degrees and is provided with reversing link motion. The double drums regulate both the hoist at a speed of 300 feet per minute and the travel along the cable at 1,000 feet per minute. A 70-horse power boiler is required.

The carriage and skip, which are automatic in action, are shown in Fig. 95, the capacity of those on the Drainage channel being 1.8 yards, and the average of a month being about 600 yards per day of ten hours. The cost of operation, including labor, fuel and everything except interest on plant

118   *THE COFFER-DAM PROCESS FOR PIERS.*

and repairs was less than $18 per day or from three to four cents per yard.

The cableway on the Coosa dam and lock (Fig. 96) had a capacity of about eight tons and made a round trip on an average of about three minutes.

FIG. 96.—LIDGERWOOD CABLEWAY AT COOSA DAM. SPAN 1012 FEET.

Such a plant is out of reach of high water and of trains where used over railroad tracks as at the North avenue bridge in Baltimore.

The Court street stone arch bridge at Rochester, N. Y., of eight spans, was constructed with the aid of a cableway, which was also used to remove the old bridge and piers. A cableway of one span was used to construct the Melan concrete arch bridge at Topeka, Kan. The bridge has five spans and a total length of 650 feet. During the extreme high water in the early part of 1897, when everything was completely inundated, and an ordinary derrick plant would have been swept away, the cableway was high and dry out of reach of the flood.

The prevailing low prices of contract work make it necessary to employ every improvement on important engineering work, and the cableway has doubtless come to stay as one of the most remarkable of our tools.

ARTICLE X.

THE COFFER-DAM PROCESS FOR PIERS.

LOCATION AND DESIGN OF PIERS.

PIERS of a bridge cannot always be located with reference to easy construction nor spaced at economical distances apart. In thickly settled parts of a country, or as part of an existing line of communication, the bridge must be located usually in a position previously determined, and the piers can only be spaced with regard to economy, provided due regard can at the same time be paid to the needs of navigation, government requirements and sufficient waterway.

Where the bridge is to be constructed in a new country, or upon a new line of road, the crossing should be selected where the river is of moderate width; that is, not so wide as to demand a structure of excessive length and probably of excessive cost, nor so narrow that the current will be exceedingly swift and make the foundations very difficult and costly to build, unless, of course, it is narrow enough to admit of using a one span structure at a reasonable cost.

On all the large navigable rivers, the channel is fixed and the length of the channel span prescribed by law, as is also the method of procedure in obtaining the approval of the government engineers. The Secretary of War must be furnished with a copy of the state law authorizing the construction of the bridge, certified to by the Secretary of State under seal; drawings in triplicate showing the general plan of the bridge; a map in triplicate showing the location of the bridge, giving for the distance of one mile above and one-half mile below the proposed location, the high and low water lines upon the banks of the stream, the direction and strength of the current at high and low water, with the soundings accurately showing the bed of the stream, and the location of any other bridge or bridges, such map to be sufficiently in detail to enable the Secretary of War to judge of the proper location of the bridge. In addition to the above, if the applicant is a corporation, there will be required a certified copy of their articles of incorporation, a certified copy of the minutes of the organization of the company, and an abstract of the minutes of the corporation, showing the present officers of the company, all duly certified to.

When the location of the bridge has been made, a thorough examination of the site must be instituted. Soundings must be made to determine the depth of the stream at low water; ordinary and extreme high water lines must be established and the flow of the stream be obtained. A careful examination must be carried out as to the character of the river bed, and drillings made to learn the character and thickness of strata and the distance to bedrock, as well as the quality of it.

Borings to a small depth may be made by hand drills (Fig. 97a), which are operated by striking with a sledge and turned constantly to keep a round hole, or if long and heavy they will cut their way, if simply raised up and

HAND DRILL AND SWAB.

allowed to drop, with their own weight. The hole is kept partly filled with water and can be cleaned out with a small sand pump or with a swab (Fig. 97b) made from a stick slivered at the end, which will also bring up samples.

The Pierce steel prospecting augur is a tool, which can also be used without a derrick to bore test holes from 10 to 50 feet into loose soils or clay. Holes from 2½ to 6 inches in diameter can be drilled and samples obtained. The augur can be turned either by hand wrenches or by horse power.

Where the borings are to be of an extensive character a well-drilling machine can be utilized, such as shown in Fig. 98, and which can be run onto an ordinary flat-boat and towed to place.

The tools for drilling are a temper screw for regulating the height of the drill, a sinker bar to give the weight, steel jars and drilling bits. A sand pump is used to clean the hole and obtain samples; rope spears, rope knives and fishing tools to remove lost rope, tools and pebbles or other obstructions. The drill holes, unless through rock, are cased with iron pipe which can be withdrawn when the hole is completed.

The borings made by the Mississippi River Commission were very extensive and a special tripod apparatus (Fig. 99) was devised with a view to easy transportation and easy repair in the field. The tripod was 30 feet in height, with a strong head or cap, surmounted by a galvanized iron guide pipe 20 feet in height, in two sections, and held in place with guy ropes. The men operating the tools stood upon the triangular platforms which were attached to the legs. The casing was iron pipe in 10-feet lengths and screwed together so as to present a smooth surface, while the bottom was provided with a steel cutting shoe, having a mouth slightly larger than the pipe. The sinking is accomplished by driving and by twisting; the driving being done by means of the clamp on the pipe and the maul sliding on the pipe. (Fig. 100.) The weight of the maul is from 80 to 100 pounds and is

FIG. 98.—STEAM POWER WELL DRILLER.

worked by three men giving it a lift of about 2 feet, the best results being obtained when the men act in concert and give rapid blows. The removal of the core and samples is accomplished by means of the various tools shown

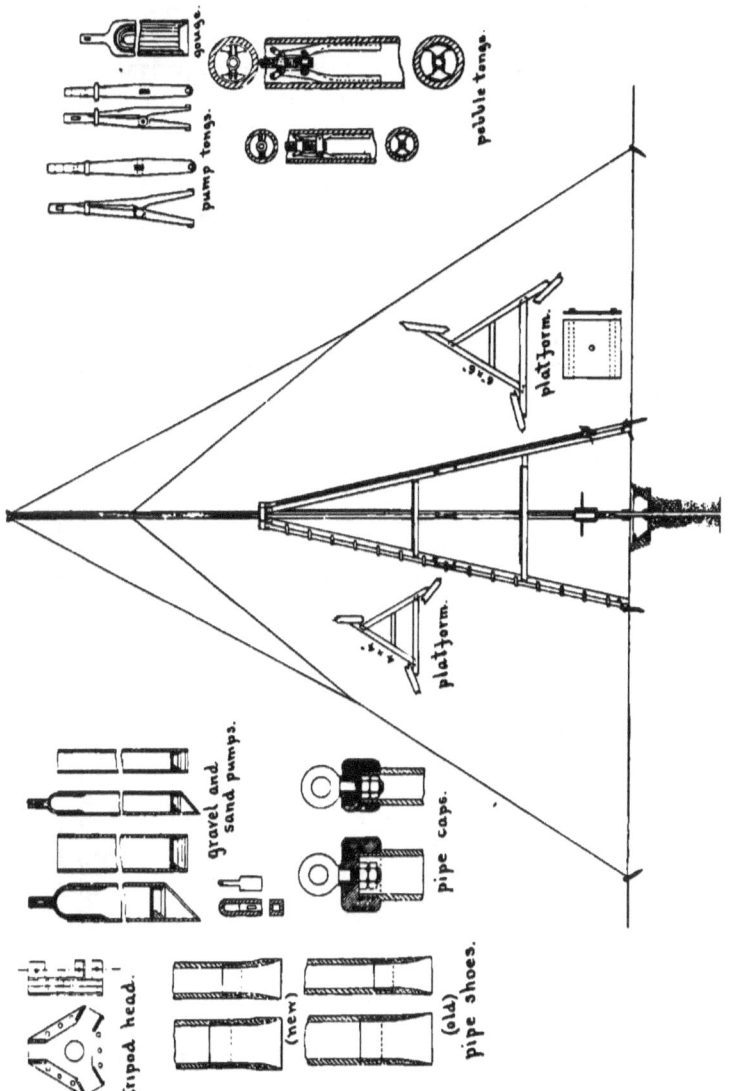

in Fig. 99, and requires great care and considerable experience. The pump was raised and lowered by means of the reel attached to one leg of the tripod, and its distance from the surface noted from graduations on the pump rod. When the boring is completed the tube is withdrawn by a system of compound levers, assisted by a set of differential blocks when necessary, as the force exerted was often as much as the strength of the pipe at the joints. The pebble tongs were for use in removing large pebbles which would not enter the pumps, and for recovering lost tools or the pump itself in case of becoming detached.

The above account is taken from the report of J. W. Nier, Assistant Engineer, to which reference must be made for other details.

When the examination of the site has been completed and the borings finished, the form of foundations may be decided upon, due weight being given to good foundations and to the allowable expenditure. Should the obtaining of good foundations be seen to be very expensive, long spans must be adopted to require few piers in the river, but if inexpensive much shorter spans, with more piers may be used.

The length of spans for a least cost of structure was formerly assumed to be decided when the cost of one span was made equal to the cost of one pier, and for spans of certain capacity this might be approximately true, but a very neat mathematical solution of this problem by Alfred D. Ottewell, Consulting Engineer, was published in the Engineering News of December 14, 1889. The total length of the structure in feet was represented by l, the number of spans by n, the length of one span in feet $l \div n$ by s, the cost of one span in dollars by c, the cost of one pier in dollars by p, the total cost of the structure in dollars by y, while a and b are constants.

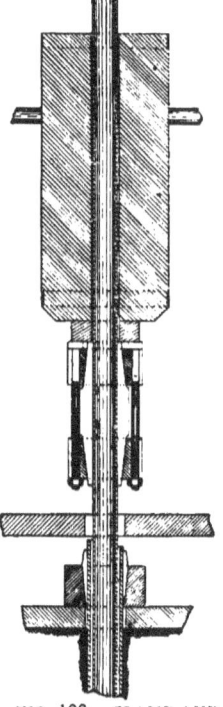

FIG. 100.—CLAMP AND MAUL.

From the estimated cost of a large number of spans, a curve of costs was plotted and the following equation of a parabola deduced:

$$c = a + \frac{(s - 20)^2}{b} \quad (1)$$

Since there are n spans and $n - 1$ piers, the total cost of the structure would be

$$y = nc + (n + 1)p \quad (2)$$

Then by substituting the value of c from (1), reducing and making the first differential coefficient equal to zero the cost of one pier is obtained, which will make the total cost of the structure a minimum or

$$p = \frac{s^2 - (ab + 400)}{b} \quad (3)$$

Or when the cost of a pier has been estimated, the economical length of span may be found by a transposition of the above formula:

$$s = \sqrt{ab + 400 + pb} \quad (4)$$

The values of a and b may be found by substituting in equation (1) computed values of the cost of a number of spans for an actual loading. Values of s, p and c, may then be computed and tabulated for spans from 100 feet upwards, as formula (1) is not true for shorter lengths.

In an actual calculation for B. & O. R. R. loading, which consists of two 125-ton engines followed by a 4,000-pound per lineal foot trainload, a was found equal to 1950 and b to 3.05. Assuming a case where the length of the bridge is 700 feet, where the height of the piers will average 25 feet, and the average cost of piers and abutments be $4,310, then from formula (4) the economical span will be found equal to 140 feet. The total cost of the structure will be found, by using formula (1), and the cost of piers as above, to be $59,700. While with only four spans of 175 feet, the total cost would exceed $60,800, and with six spans of 117 feet, would be about $61,400.

Should there be any doubt as to the ease of obtaining foundations, the prudent engineer might deem it wise, however, to build the four-span structure and avoid the risk and delay which would be caused by another foundation in the river.

After deciding upon the number and location of the piers, they must be designed with reference both to their being as slight obstructions to the water as possible and to their architectural appearance.

The design of piers has been given particular attention by Geo. S. Morison, Consulting Engineer, whose work on the bridges across our great rivers is notable for its strength, simplicity and finished appearance. In a recent lecture he describes the process of the design of some large piers: "Fourteen years ago I had occasion to design a bridge pier for a bridge across one of our Western rivers, and I tried to make an ornamental pier. When the plans were completed I did not like them. One change after another was made, all tending to simplicity. Finally the plans were done. From high water down, the pier was adapted to pass the water with the least disturbance; it had parallel sides and the ends were formed of two circular arcs meeting. Above high water the ends were made semi-circular instead of being pointed. The pier was built throughout with a batter of one in twenty-four. A coping two feet wider than the body of the pier projected far enough to shed water, and the projection was divided between

FIG. 101.—PIER OF OMAHA BRIDGE, UNION PACIFIC.

the coping and the course below. Another coping with a less projection surmounted the pointed ends where the shape was changed. It was as simple a pier as could be built, and in every way fitted to do its duty. I had started to make a handsome pier. The pier that was exactly what was wanted for the work, was the only one that satisfied the demands of beauty. Forty-three piers of precisely this design (no change having been made except in the varying dimensions required for different structures), besides eight others in which only the lower parts are modified, are now standing in eleven different bridges across three great Western rivers. In designing a pier it must be remembered that the portion of the pier below the water has more to do with the free passage of the water than that above water. In a deep river the model form of the pier should begin near the bottom of the river and not at low water. Many rivers in flood time carry a great amount of drift. A pier like that which I have described catches but little of this drift. If, however, a rectangular foundation terminates but little below water, that foundation may both disturb the current and catch the drift."

FIG. 102.—RUSSIAN PIER, RUSSIAN STATE RAILWAYS.

The piers of the Omaha bridge, which carries the Union Pacific across the Missouri river, are illustrated in Fig. 101, and were constructed as described and are among the most beautiful piers in this country.

In Europe, where money is more lavishly expended on great works of engineering, piers of great architectural beauty are much more frequent. The Russian Government railways, which have seemingly been constructed without regard to expense, have many beautiful examples of bridge masonry and piers; the view of one of them (Fig. 102) with curved ends, shows the elegant and massive character of the masonry. While extremely

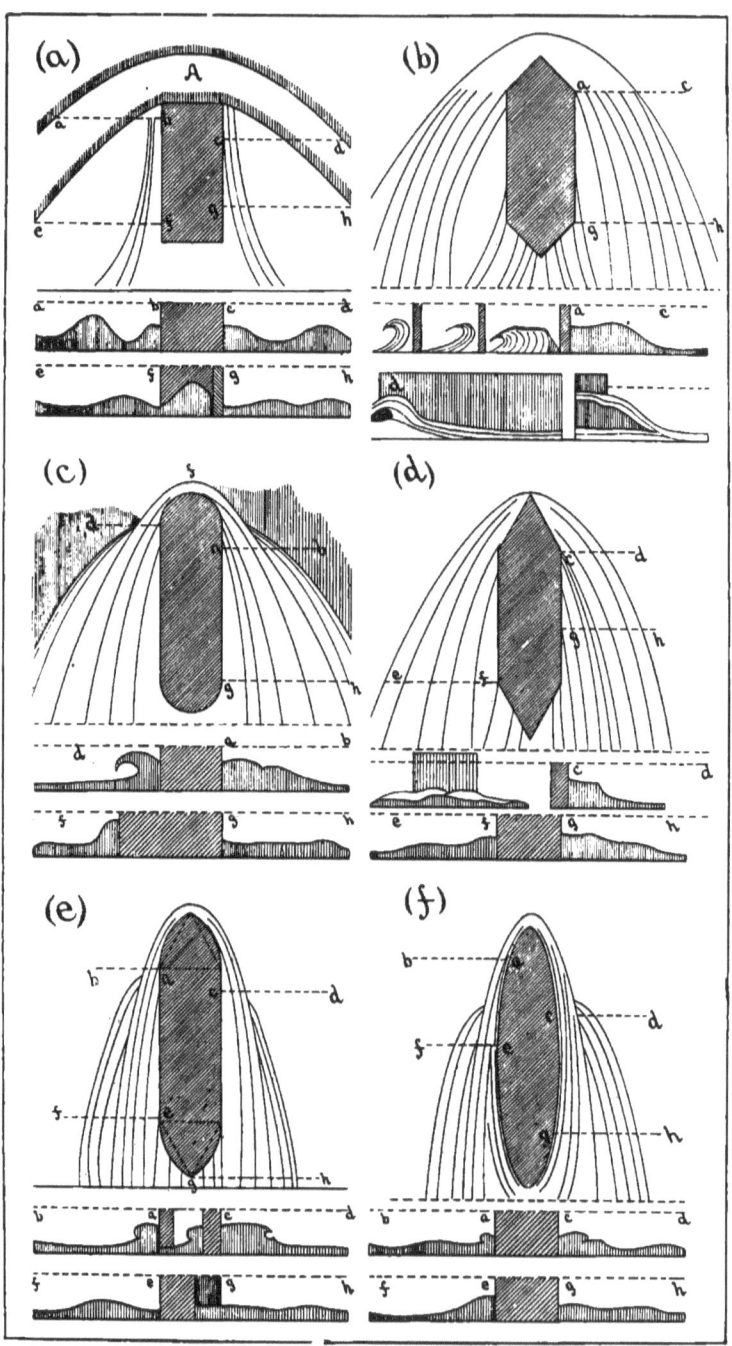

FIG. 103.—CRESY'S EXPERIMENTS ON THE FORM OF PIERS.

simple in design, the cut stone coping and the moulded corbel course below give it a finish which cannot be surpassed.

The design of piers for strength and stability is fully treated in Baker's Masonry Construction, but some experiments, which were made with reference to the proper form to occasion the least resistance, will be quoted at length from Cresy.

The introduction of piers into a channel gives rise to a great disturbance in the velocity and flow of the water. Rapid currents are formed which cause the bed of the stream to become washed and the foundations to be endangered; eddies are created which are likewise undesirable, and it becomes necessary to adopt such a form for the ends of the piers that the disturbance to the flow shall be small.

M. Bossut, in a French work on jetties, thought to have solved this problem by mathemathics, his conclusion being that the starling should be triangular, the nose being a right angle.

M. Dubuat, in his "Principles of Hydraulics," gave another solution which was more nearly the truth, in that he arrived at the conclusion that the faces of the starling should be convex cnrves. The true form is most nearly reached when these curves are tangent to the sides of the pier, and further than this, regard must be paid to giving enough solidity to the starlings to protect them from ice and drift. A happy medium would seem to be reached, by making the curves with a radius equal to one-sixth of the circumference, described on the sides of an equilateral triangle.

Experiments were made with models of different forms, which were placed in a rectangular canal between boards of 50 centimeters in length, in which the water flowed about 40 millimeters in height, the models being 15 centimeters in thickness. By means of a fall, the water was given a velocity of 3 meters 9 centimeters per second, the contraction, eddies and currents being carefully measured. The first experiment was made on a pier (Fig. 103A) with rectangular starling. An eddy was formed before the pier 34 millimeters high, in a nearly circular band A, falling nearly vertical at the corner. There were two other currents along the faces of the pier, the height of which can be seen in the cross-sections.

The second experiment (Fig. 103B) was with a triangular starling, the nose being a right angle. It formed a less obstruction than the square end, but the fall at the shoulder was as deep and more dangerous, while eddies were formed as seen in the sections.

The third one (Fig. 103C) had a semi-circular starling. The eddy was not so wide, but nearly as high.

The fourth model had a triangular starling, with an angle of 60 degrees at the nose. (Fig. 103D.) The eddy was less, as was also the fall at the shoulder.

The starling in the fifth was formed by two circular arcs, tangent to the sides and described on the sides of an equilateral triangle. (Fig. 103E.) The eddy was small and there was no fall at the shoulder.

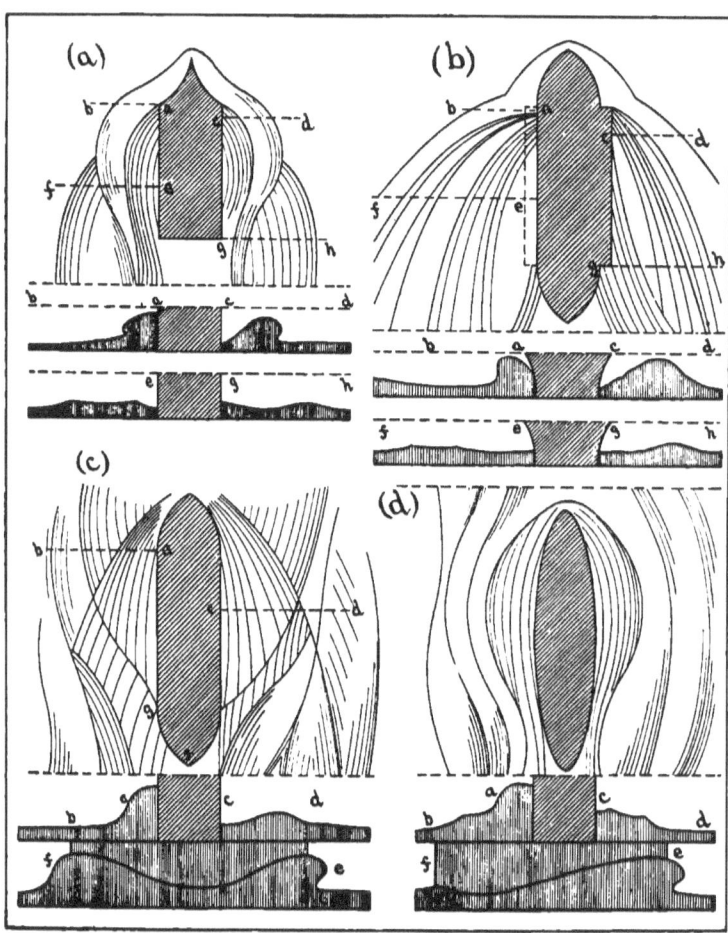

FIG. 104.—CRESY'S EXPERIMENTS ON THE FORM OF PIERS.

The sixth (Fig. 103F) was a model, the plan of which was an ellipse, of which the small diameter was one-fourth the length, and the eddy was less than any of the others.

The seventh model (Fig. 104A) had a starling with concave faces, such as is sometimes used where the wing wall meets an abutment. It produced the most dangerous currents of all.

The eighth (Fig. 104B) was of the same form as Fig. 103E, but the water was supposed to mount the springing of the arch.

The ninth and tenth experiments (Figs. 104C and 104D) were on the same forms as Figs. 103E and 103F, but the current had a velocity of 4 meters 87 centimeters per second, such as a river would have in its overflow. The eddy (Fig. 104C) rose to nearly twice the height, as was the case with the lesser velocity, and while there was no fall, the inclination formed along the faces was more rapid.

The effect with this velocity on the elliptical pier (Fig. 104D) was similar to the lesser velocity but more marked. It may thus be concluded that the elliptical section offers the least resistance to the current and occasions the least contraction, while the form with convex starling comes next, and of piers with triangular starlings the one with the 60-degree nose is the best.

Where ice is to be provided for, the nose is often inclined to allow large cakes to mount it and break in two, without doing further damage. For any large or important structure, the design of the piers should receive a great deal of study, and be designed not only with reference to their theoretical form, but with reference to the form of pier which has shown the best results practically and has been found to be best suited to the velocity of the stream in which they are to be built, and to best withstand the drift and ice that may be met with, giving at the same time all the consideration possible to the architectural effect and to the harmony with the entire structure.

APPENDIX.

SELECTIONS FROM SPECIFICATIONS.

SPECIFICATIONS FOR COFFER-DAMS AND FOUNDATIONS, OHIO RIVER MOVABLE DAMS.

MAJOR W. H. HEUER, U. S. Engineer.

GENERAL DESCRIPTION.

The site of Dam No. 2 is on the Ohio River, distant from Pittsburg, Pa., $10\frac{1}{4}$ miles, and adjacent on the right bank to the Pittsburg, Ft. Wayne and Chicago Railway. It has Neville Island on the left bank, and is accessible by street cars from Pittsburg.

The lock is to be located on the left bank of the Ohio River, immediately behind Merrimans dyke. It will be in general dimensions the same as locks Nos. 1 and 6, viz., 110 feet wide and 600 feet long.

SPECIAL DESCRIPTION.

The river bed at No. 2 consists of gravel throughout, and the excavations will be made to a depth sufficient to insure a permanent and enduring foundation, which will ordinarily be 14 feet below the gate sill, but may be otherwise, as the Engineer, in his judgment, may direct.

The work will conform to the drawings exhibited, and to such others, in explanation of details or modifications of plans, as may be furnished from time to time during construction.

CONTRACTOR TO FURNISH ALL MATERIAL AND WORK.—It is understood and agreed that the contractor, under his contract prices for work in place, is to furnish and pay for all materials, stone, cement, sand, earth, timber, material for coffer-dam and protection cribs, excavation, lock-filling and discharging valves (set in masonry), flushing valves, anchor bolts, lock-gate tracks, and everything entering into or connected with either the permanent or temporary construction, and he is also to supply and pay for all work, skilled and otherwise, required to prepare and place the materials, and complete the work according to the drawings and these specifications.

CONTRACT TO INCLUDE.—The contract will cover the construction and completion of the foundation, masonry and timber work of the lock, including both land and river walls, the gate-recess walls, the foundations of the lock-gate tracks, the guiding walls above and below the lock, the pipe and flushing conduits, the drift chute, the foundations for the power-house and lock-keepers' residence, and every such other permanent construction as shall be shown upon the drawings. It shall also include the clearing of the land necessary for the proper execution of the work embraced, in this contract, all pumping and bailing, dredging and excavation, puddling and embankment, the construction of all coffer-dams, stone masonry, concrete and brick masonry, timber work and iron work, and all such other work which, in the judgment of the Engineer, is necessary and included in the proper completion of the contract.

TOOLS, MACHINERY, BUILDINGS, ETC.—The contractor, without cost to the United States, shall furnish all appliances, dredges, pumps and pumping machinery, boats, tools, derricks, tramways, foot walks, roads and landings, and all needful temporary buildings and shops.

COFFER-DAM.

SHEETING.—The coffer-dam, about 1500 feet in length, shall be built as shown generally by the drawings exhibited, and as directed by the Engineer. It shall be 14 feet high above the sill of the lock, and shall consist of two walls or rows of plank sheeting, spaced 12 feet apart in the clear, driven or set firmly from 1 to 2 feet into the river bed, and supported laterally by horizontal longitudinal stringers, the latter being spaced at varying intervals, increasing in width from the bottom to the top, and to be sufficiently and firmly bolted together transversely with iron rods passing through the coffer-dam horizontally from the rows of stringers on the one side to the corresponding rows on the other, against which the vertical plank sheeting shall be securely spiked.

FILLING AND DECKING.—The interior, or space between the walls of sheeting, shall be filled with heavy dredged river-bed or other material not liable to wash, and to be covered over with a suitable decking of plank (to protect it from injury in case of being submerged by floods), all complete as shown on the drawings.

PILING AND CRIBS TO PROTECT.—At the upper outer corner of the coffer-dam shall be placed a crib built of framing timber and filled with riprap stone ; from the upper corner of the crib, at an angle of 45° with the axis of the current, a line of piling, spaced 5 feet apart, firmly bolted together with waling pieces, shall be driven to the shore to form a protection to the coffer-dam ; also outside and along the coffer-dam, from the upper outer corner to the lower corner, clusters of piles, firmly bound or bolted together, shall be driven at intervals of about 80 feet. The tops of all piling shall be sawed off to a uniform height of 2 feet above the coffer-dam. Protection cribs shall be placed at such other points along the coffer-dam as may be shown upon drawings.

HOW PAID FOR.—Bidders will state a price per lineal foot of coffer-dam completed. No payment will be made for any portion thereof until the entire coffer-dam is completed. Drawings will be furnished, showing the general type of the coffer-dam and its manner of construction, and every detail necessary for intelligent bidding. Should any work on the outside of the coffer be necessary, such as gravel filling or riprapping, it shall be paid for at the price bid for gravel filling, riprapping, etc. If, owing to the nature of the river bed, it shall be found impossible to drive the plank sheeting to the required depth, then the contractor, after driving the sheeting as deep as possible without injury, and in lieu of driving it to its full depth, shall fill around the outside of the walls with the same material as is used in filling the coffer-dam, to the height of 4 feet above the surface of the river bed, and for which no extra compensation will be allowed.

REMOVAL OF.—The contractor will be required to remove the coffer-dam and its belongings at his own cost. The time and manner of the removal of the coffer-dam, or any part thereof, and the place to deposit the materials, shall be prescribed by the Engineer.

TO BELONG TO THE UNITED STATES.—It is understood and agreed that the payments made for the coffer-dam, including the crib and pile protection, shall cover the entire cost thereof to the United States, and by virtue thereof they shall become the property of the United States. The contractor, however, must maintain the same and make all needed repairs to same during the existence of the contract, without expense to the United States.

DEPOSIT WITHIN THE COFFER-DAM.—Material washed or left in the space enclosed by the coffer-dam by freshets shall be removed by the contractor, as directed, at his price for common excavation, which price shall cover all necessary cleaning and scrubbing. No payment will be made, however, for removing material washed into the enclosure from the coffer-dam itself or from any deposit made by the contractor on or above the works.

MATERIAL AND WORKMANSHIP.

TEMPORARY PILING shall include all piles driven for the protection of the coffer-dam and "deadmen" for derricks. They shall be of good quality, round oak timber, not less than 12" diameter at the butt, and of length varying from 20 to 25 feet, and longer if necessary.

SHEET PILING.—In excavating for foundations, should quick-sand or fine-sand-carrying water be encountered, close sheet piling will be required to be driven to whatever extent the Engineer may direct.

SHEETING.—The sheeting shall include the walls and decking of the coffer-dam, including the stringers; also such shoring as may be directed by the Engineer to remain in the finished structure. It shall consist of the best quality of hemlock obtainable, and must be in all cases satisfactory to the Engineer in charge.

GRAVEL OR EARTH FILLING.—Gravel or earth filling will include all material used in filling the land-wall enclosure, back of the guiding walls, etc. It does not include any filling in the construction of the coffer-dam.

STONE FILLING shall include all stone placed in the protection cribs or any riprap stone ordered for the protection of the work.

CRIB WORK shall be built of hemlock framing timber framed together in square bins and securely bolted together by iron drift-bolts. The interior of the cribs shall be filled with riprap stone, and should the Engineer deem it necessary such riprap stone shall be placed on the outside of the crib. The whole to be built as shown by the drawings.

FRAMING TIMBER.—For all temporary crib work, also the permanent crib at the head of the upper guiding wall, framing timber shall be used. No stick shall be less than 10" × 10" in section.

"Framing timber" is a commercial term for a class of timber hewn to various sizes.

EXCAVATION.

TO INCLUDE.—It shall include the removal of all gravel or other material to the depth required for the lock and its upper and lower entrances, the gate recesses, Poiree-dam and gate-track foundations, for the foundations of all walls, and for all conduits or wells, and all such other material as may be found necessary in the judgment of the Engineer to be removed for foundations and otherwise in permanent construction. It will include all dredging and all material excavated of whatever nature, however removed, for foundations and for site of coffer-dam.

LINES, SLOPES, AND GRADES FOR.—All excavations shall conform to such lines, slopes, and grades as may be given by the Engineer, and anything taken out beyond such given limits will not be paid for by the United States.

MATERIAL TO BE DEPOSITED.—Excavated material is to be deposited as and where directed by the Engineer. It shall be deposited in such manner as not to interfere with present or proposed navigation. Material of any kind deposited by the contractor in absence of, or in disregard of, instructions, shall, if required by the Engineer, be removed by the contractor at his own cost.

SHORING.—All excavation for foundation shall be securely shored and thus maintained until the foundation has been sufficiently advanced to dispense with the same, when it may remain or be removed at the discretion of the Engineer.

DREDGES AND PUMPS.—The contractor will be required to employ, at the same time, not less than two suitable steam dredges at excavating and filling; and for pumping he must keep at least three good sufficient pumping outfits, with pumps, engines, and boats complete, in or always ready for operation. The dredges must be equipped to do effective work to a depth of 28 feet.

FOUNDATIONS.

CHANGES OR MODIFICATIONS OF.—The character of the river bed and of the proposed foundations for the different parts of the work is shown in general on the drawings and cross-sections exhibited, and it is understood that the United States shall have the power to make any changes in the plans of the foundations that may, in the judgment of the Engineer, be considered advisable after examinations made, as the excavations proceed within the coffer-dam after it is pumped out and it is understood and agreed that the contractor shall have or make no claim against the United States on account of any such changes in or modifications of the plans of the foundations, or on account of any increase or decrease in the depth of same, under or over those referred to herein or shown on the drawings exhibited.

MASONRY.

CEMENT.—Cement will be of uniform quality, setting well both in air and water, and free from anything that will cause the mortar to swell, crack, or scale. It shall be put up in strong, sound barrels, well lined with paper so as to be reasonably protected from air and moisture. The average net weight of the barrels shall be not less than 265 pounds, unless expressly so stated in the proposal. Each barrel must be labeled with the name of the brand and of the manufacturer.

In general, ten barrels of every one hundred will be tested.

The cement must stand the following tests: Fineness—At least 85 per cent must pass a sieve of 6400 meshes to the inch. Setting—Cement must be moderately slow setting; it must not begin to set within fifteen minutes, as determined by Vicat needle 1/12 inch in diameter with 1/4 pound load, and it shall not bear the weight of one pound on wire 1/24 inch in diameter within thirty minutes, but must bear such weight within one hour and a half. Strength—The minimum tensile strength per square inch of briquettes of neat cement mixed with about 33 per cent of water by weight, and exposed in air for one hour, and the remainder of 24 hours in water, shall be not less than 50 pounds; with longer time, whether in air or water, there must be a decided increase of strength; it must also test to the satisfaction of the Engineer when mixed with sand. The tests for setting will be made at a temperature of air and water of about 75° Fahrenheit. All other tests will be made at a temperature above 60° Fahrenheit. The cement will be subject to inspection at all times, and must be kept well housed.

SAND.—The sand used must be clean, sharp, washed, river sand, satisfactory to the Engineer.

MORTAR.—To be composed generally of two parts of sand to one of cement; when required, and whenever thought necessary by the Engineer, it shall be made richer. It must be thoroughly mixed and used before it has begun to set. If required by the Engineer, the mortar beds will be protected from the sun.

POINTING.—All face work is to be pointed, as fast as the work progresses, with stiff mortar, mixed, one of sand to one of Portland cement, thoroughly hammered in and finished with proper tools; before the final acceptance of the work all face masonry which at that time does not appear properly pointed shall be repointed by the contractor to the satisfaction of the Engineer, without extra cost.

FROST.—Masonry will not be executed during freezing weather, nor when, in the judgment of the Engineer or his agent, it is likely to freeze before the mortar shall set. To guard against injury from frost all new and unfinished work shall be properly protected by the contractor at his own cost.

VOIDS AND OPENINGS.—Due regard shall be had in the construction of all masonry walls to leave all necessary voids or openings for conduits or wells, or for such other purposes as may be required by the Engineer.

ASHLAR.—It shall comprise such part of the walls as is built of stone, with point-dressed face, and beds and joints smoothly and squarely dressed.

QUALITY OF STONE.—All stone shall be perfectly sound, strong, hard, free from injurious seams, and in all respects satisfactory to the Engineer. Stone to be such as can be truly wrought to such lines and surfaces, whether curved or plain, as may be required. No stone shall be used which weighs less than 135 pounds to the cubic foot.

SAMPLES OF STONE.—Each bidder must deposit at this office, all charges prepaid, before the bids are opened, a 6-inch cubical block of the stone he proposes to furnish, and state the quarry from which it was obtained. The quality of the stone must be at least equal to that of the sample. The sample must be truly squared, and dressed on four sides; one side must be hammer-dressed, one side smooth-dressed and rubbed, and one side pitch-dressed. The other two sides are to be left with quarry face.

STONE MAY BE REJECTED.—The United States reserves the right to reject any stone not deemed suitable, or which, during the execution of the contract, shall be found defective. The beds of the stone must be their natural quarry beds. No lewis or dog holes, letters, or marks of any kind will be allowed on any dressed face of stone, but each face shall have left on it a boss for lifting, to be removed by the contractor after the stone has been set.

DRESSING OF STONE.—Stone must be accurately cut, square and true, and the faces must be pitch draughted and point-dressed to a plane with the draught, forming an approximately smooth surface. The beds must be smoothly and squarely dressed, full length and width. The vertical joints must be dressed to a depth of not less than 18 inches from the face, and the allowance for joints must not exceed 3/8 inch. One-third of the stone in each course must be headers. All stones not accurately dressed will be rejected. All dressed stone must have the dimensions plainly marked on one end.

DIMENSIONS.—The cut-stone stretchers must be not less than 3 feet nor more than 5 feet long, and their width must be not less than 1½ times the height of the course to which they belong. The width of the headers must be not less than 1½ times their height, and their length must be at least double their breadth, unless otherwise ordered. The thickness of courses includes the joint, which will be 3/8 inch.

LAYING STONE MASONRY.—The faces of the walls shall be accurately laid to the lines indicated on the drawings, or as directed by the Engineer. All stones to be well laid to proper lines, in full beds of mortar, and settled in place with a wooden maul; the use of grout is prohibited. No dressing, except in special cases, and by permission of the Engineer, will be allowed on backing after it is laid in the wall. The bond of stone shall in no case be less than 9 inches. The walls will be laid in horizontal courses throughout, each course to be of uniform height through the wall. Heights and arrangements of courses to be determined by the Engineer. When laying masonry the site for the stone shall be thoroughly cleaned with a scrub broom and moistened; and the stone shall always be cleaned and well moistened before being set. Not more than three unfinished courses of face stone will be permitted upon the wall at the same time, without special permission from the Engineer in each case. Proper machinery must be used in handling the stone, face stone shall not be disfigured by use of plug or grabs. Any stone chipped or spalled shall be rejected. Stones having defects concealed by cement or otherwise will be rejected on that account alone.

COPING.—The coping will be of the same class and quality of stone described in ashlar masonry. It will be carefully and truly cut to forms and dimensions given, from the best stone; it will be crandalled on all outer faces; the exposed edges of the coping to be rounded to a radius of 3 inches and chiseled smooth where required. Beds and vertical joints to be pointed true and full throughout and be laid with 3/8-inch joints.

The coping is to be doweled as required by the Engineer with round iron. The dowels to be furnished and placed by the contractor. The drilling for and placing of the dowels will be covered by the price for "Bolt Holes in Masonry." The dowels will be set in Portland cement.

RUBBLE STONE.

QUALITY AND DIMENSIONS OF.—Rubble stone must be sound, hard, and durable, free from seams, scale, earthy matter, and other defects. Rubble stone shall in general be not less than 3/4 of a cubic foot in size. It must be in fair shape for laying in the face of the walls without dressing. No spalls will be allowed.

LAYING.—The stone must be laid on their natural bed in full beds of hydraulic cement mortar, with all joints and voids well filled with mortar. Leveling up under stones with small chips or spalls will not be allowed.

The stone shall be carefully selected for the outer face so as to have vertical joints and present a good face of broken rough masonry.

CONCRETE.

COMPOSITION OF.—Concrete shall be composed of satisfactory cement and river gravel; the latter, should it be of an approved quality, shall be taken from the various excavations of the lock and its walls. This gravel generally has a sufficient volume of sand to fill all voids, should there be a deficiency of sand in any portion of the gravel the contractor will be required to supply said deficiency by good, sharp, washed, river sand. The quantity of cement to be used will generally be about 20 per cent greater than the volume of voids in sand and gravel.

MIXING AND PLACING OF.—The concrete is to be well and rapidly mixed by machinery, as may be required by the Engineer, unless otherwise specified. It will be deposited in layers not more than 8 inches thick; wherever and whenever required, the layers will be thinner than 8 inches, and all thoroughly rammed by such process as the Engineer may approve.

RIVER WALL.—In the river wall of the lock the concrete shall be laid in courses of a thickness corresponding to the adjoining courses of ashlar masonry. It shall be filled in flush with the top of each course before the next course of ashlar above shall be laid.

Before putting in the concrete of any course the bed and adjoining course of ashlar shall be thoroughly wetted so that no dry surface may come in contact with the fresh concrete, destroying its power of adhesion by absorbing its moisture.

In order that the work once begun may progress without delay all cut stone needed for the ashlar facing shall be on the ground when the concrete foundation has been completed.

TIMBER IN PERMANENT CONSTRUCTION.

TO CONSIST OF all timber used in the timber facing of the lock walls and the guide walls; all timber cribbing in the gate-track and Poiree dam foundations; the oak sheeting at the head of the guide walls; and such other timber in permanent construction as shall be shown upon the drawings.

GENERAL QUALITY AND DIMENSIONS.—All timber must be first class, and any of inferior quality will be rejected. Sap-wood in any stick will cause its rejection. The timber must be free from black or rotten knots, wane edges, wind shakes, dose, or other imperfections. Firm sound knots, if not too numerous, will not be considered defects. Timber must be full to size, true, and out of wind, and when required must be sawed large enough to dress down to required dimensions. The timber will be inspected on arrival at the work, and if found to be defective will be rejected.

OAK.—Oak timber must be taken from the best quality live white oak sawed timber.

WHITE PINE.—Shall consist of the best quality of clear white pine obtainable.

HEMLOCK.—Shall be the best quality of hemlock obtainable.

FRAMING, ASSEMBLING, AND PAINTING.—All timber must be accurately framed, fitted, and assembled, according to detailed drawings and directions. As the timber is framed it shall be painted about the ends and elsewhere as may be required to prevent checking. The paints for this will be furnished and applied by the contractor, and covered in his price for " Timber in Permanent Construction."

TIMBER FACING, UPRIGHTS, AND SHEETING shall be constructed of oak, and shall consist of uprights spaced at intervals of 6 feet, center to center, anchored to the concrete masonry by tee-head screw bolts as shown on drawings. To the uprights shall be bolted, with wrought-iron screw bolts, oak sheeting 6 inches thick.

NOSING TIMBER shall extend along the top of the guide wall, forming a cap to the uprights and securely bolted to them, as shown on the drawings. The top surface of the nosing shall be flush with the top of the concrete masonry wall.

OAK SHEETING.—This refers to the sheeting on the upper faces of the protection crib for the upper guiding wall at the upper end thereof. It shall be spiked on and firmly held in place with iron bands or straps bolted to the framing timbers of the crib, if, in the judgment of the Engineer, this may be deemed necessary.

SUPERVISION AND MEASUREMENT OF WORK.

INSPECTION, REJECTED MATERIAL, ETC.—The works will be conducted under the direction of the local or resident Engineer, who shall have power to prescribe the order and manner of executing the same in all its parts; of inspecting and rejecting materials, work, and workmanship which, in his judgment, do not conform to the drawings that may be furnished from time to time, or to these specifications. And any material, work, or workmanship so rejected by him shall be kept out of or removed from the finished work, and no estimate or payment shall be made until such material, work, or workmanship be so removed.

When so required rejected material shall be piled up in sight near the works and kept there until the Engineer gives permission to have it removed.

The United States will keep inspectors on the work who will receive instructions from the resident Engineer. They will have power to object to any materials, work, or workmanship. Any material, work, or workmanship objected to by the inspectors shall be kept out of or removed from the finished work, unless in each particular case the objections of the inspector shall be overruled by the local or resident Engineer; and, unless the objection be so overruled, no estimate or payment shall be made until such material, work, or workmanship be so removed.

The local or resident Engineer shall have power to overrule or rescind any or all objections or decisions of the inspector.

The decision of the United States Engineer Officer in charge of the works shall be final and conclusive upon all matters relating to the work and upon all questions arising out of these specifications, and from his decision there shall be no appeal.

FAILURE TO PROSECUTE OR PROTECT WORKS.—If at any time the contractor shall refuse or fail to prosecute the work or provide for carrying on the same as directed by the Engineer, or fail to properly protect any part of the work, permanent or temporary, the Engineer shall have power to employ men, to purchase or otherwise provide materials, tools, machinery, etc., and put the work in proper advancement or condition, and the entire cost of so doing shall be deducted from payments to be made under this contract.

COMPLETE WORK REQUIRED.—The contractor is not to take advantage of any omissions of details in drawings or specifications, or errors in either, but he will be required to do everything which may be necessary to carry out the contract in good faith, which contemplates everything complete, in good working order, of good material, with accurate workmanship, skillfully fitted and properly connected and put together. Any point not clearly understood is to be referred to the Engineer for decision.

CHANGES.—Should any changes in the details of the shape, arrangement, or fitting of the parts be deemed necessary or advisable in the progress of the work, they must be made by the contractor, and a fair allowance will be paid for any changes which, in the judgment of the Engineer in charge, materially increases the cost of the work.

MEASUREMENT.—Measurement of all work and material shall be made in place, unless otherwise specified.

COFFER-DAM.—The price per lineal foot of coffer-dam shall include all material, lumber, iron, and gravel entering into its construction. A profile of the location will be furnished, showing a section of the river bed over which the coffer-dam is located, so that the contractor may estimate the amount of each kind of material required.

PILING.—Temporary piling shall be measured in lineal feet, and measurement shall be allowed for total length of piling used.

SHEETING —This will include all lumber used for temporary purposes, in shoring of excavations, or for forms necessary to sustain any concrete masonry until it has become sufficiently hardened. Sheeting required by the Engineer to remain in the finished structure shall be paid for at the contractor's price per thousand feet B. M. All temporary sheeting not remaining in the finished structure shall be included in the contractor's unit price for material in place, and no estimate will be made thereof by the Engineer. Coffer-dam sheeting will be included in the contractor's price per lineal foot of coffer-dam.

FILLING.—Gravel filling will be measured in the fill, and will not include any filling placed in the coffer-dam as coffer-dam filling.

Stone filling shall include all riprap work, either temporary or permanent.

EXCAVATION.—Excavation will be measured in excavation by cross-sections.

MASONRY.—All masonry, ashlar, rubble, brick, concrete, etc., will be measured by the cubic yard in place. Prices for masonry will include all required pointing. No payment will be allowed for voids or openings.

BOLT HOLES.—All holes drilled in rock or concrete or other masonry will be measured by the running foot as drilled.

TIMBER IN PERMANENT CONSTRUCTION.—Timber in permanent construction will include all timber used in any part of the permanent construction; unless otherwise particularly specified, will be classed under the following heads:

    Oak in Permanent Construction.
    Pine in Permanent Construction.
    Hemlock in Permanent Construction.

EXTRACTS FROM TOPEKA (KANSAS) MELAN ARCH BRIDGE SPECIFICATIONS.

By permission of EDWIN THACHER, M. Am. Soc. C. E.

PILING IN PERMANENT WORK.

Piling and lumber for coffer-dams to be sound white oak, yellow pine, or other woods equally good for the purpose, the quality to be acceptable to the superintendent. The piles shall be straight-grained, trimmed close, and have all bark taken off, and shall be at least 10 inches in diameter at the small end and 14 inches in diameter at the butt when sawed off. The heads shall be cut off squarely at right angles to the axis of the pile, and all piles shall be fitted to and driven with a cast-iron head. The piles shall be driven with a hammer weighing not less than two thousand two hundred and fifty (2250) pounds, and until they do not move more than three-eighths (3/8) of an inch under a blow of the hammer falling twenty-five (25) feet. No pile shall be driven less than twenty-six (26) feet below low water, and if necessary to attain this minimum depth jets shall be used in addition to hammer. The number and arrangement of the piles for each foundation are shown on the plans, and must be carefully carried out by the contractor. The piles shall be cut off at an elevation of about six (6) inches below low water. A slight variation will be allowed, but no piles must be cut off at a higher elevation. Inspection of piling and lumber, except at bridge site, shall be at contractor's expense.

COFFER-DAMS.

After the bearing piles have been driven, a permanent coffer-dam, of the dimensions marked on the plans, of Wakefield (or other equally satisfactory) sheet piling, shall be used around each foundation. The earth inside thereof shall be excavated to the depth shown on plans and replaced with concrete as hereinafter specified. During the placing of the concrete the water shall be kept out of the coffer-dams unless the bottom is so porous that it is impracticable in the opinion of the superintendent to do so—in which case some of the concrete may be placed in position by means of chutes under the direction of the superintendent until the bottom is well calked, after which the water shall be pumped out and the remaining concrete placed in position. The contractor will be required to make the sides and ends of the coffer-dams water-tight, and no leak through them will be considered sufficient cause to require any concrete to be placed by means of chutes.

CENTERING.

The contractor shall build an unyielding falsework, or centering, of the form and dimensions shown on the plans; particular care must be taken to drive the piles supporting it to a solid bearing. The estimated load upon each of these piles is twenty (20) tons. The contractor must, however, satisfy himself as to the load each pile will have to bear, and as to its supporting power. In case of any settlement the contractor shall take down and rebuild the centering and arch. The lagging shall be dressed on both edges to a uniform size so that when laid it will present a smooth surface, and this surface shall be built at the proper elevation to allow for settlement of arch, so that when the centering is struck the arch ring will come to the elevations shown on plans.

The top surface of the lagging shall be covered with W. Field's Building Paper of medium weight, known as Double Saturated Water-proof Oiled Sheathing Paper (or other equally good) to prevent the concrete from adhering thereto. No center shall be struck until at least twenty-eight (28) days after the completion of the arch. Great care shall be used in lowering the centers so as not to throw undue strains upon the arches, nor shall any center be struck before the adjoining arch has been completed for a sufficiently long time, in the opinion of the superintendent, to be uninjured thereby.

NOTE.—For the above reasons it is probable that the five centers will be in use at the same time.

PORTLAND CEMENT.

The Portland cement shall be a true Portland cement, made by calcining a proper mixture of calcareous and clayey earths; and the contractor shall furnish one or more certified statements of the chemical composition of the cement and of the raw materials from which it is manufactured. Only one brand of Portland cement shall be used on the work, except with permission of the superintendent, and it shall in no case contain more than two (2) per cent of magnesia in any form.

The fineness of the cement shall be such that at least 98 per cent shall pass through a standard brass cloth sieve of 74 meshes per linear inch, and at least 95 per cent shall pass through a sieve of 100 meshes per linear inch.

Samples for testing may be taken from each and every barrel delivered as superintendent may direct. Tensile tests will be made on specimens prepared and maintained, until tested, at a temperature of not less than 60 degrees Fahrenheit. Each specimen shall have an area of one square inch at the breaking section, and after being allowed to harden in moist air for twenty-four hours shall be immersed and retained under water until tested.

The sand used in preparing the test specimens shall be clean, sharp, crushed quartz, retained on a sieve of 30 meshes per linear inch and passed through a sieve of 20 meshes per linear inch, and shall be furnished by contractor.

No more than 23 to 27 per cent of water by weight shall be used in preparing the test specimens of neat cement, and in making the test specimens one of cement to three of sand, no more than 11 or 12 per cent of water by weight shall be used.

Specimens prepared from neat cement shall after seven days develop a tensile strength not less than 400 pounds per square inch. Specimens prepared from a mixture of one part cement and three parts sand (parts by weight) shall after seven days develop a tensile strength of not less than 140 pounds per square inch, and after twenty-eight days not less than 200 pounds per square inch. Specimens prepared from a mixture of one part cement and three parts sand (parts by weight) and immersed, after twenty-four hours, in water to be maintained at 176 degrees Fahrenheit, shall not swell nor crack, and shall after seven days develop a tensile strength of not less than 140 pounds per square inch.

Cement mixed neat with about 27 per cent of water, to form a stiff paste, shall, after 30 minutes, be appreciably indented by the end of a wire one-twelfth inch in diameter, loaded to weigh one-quarter pound.

Cement made into thin cakes on glass plates shall not crack, scale, or warp under the following treatment: three pats shall be made and allowed to harden in moist air at from 60 to 70 degrees Fahrenheit; one of these shall be subjected to water vapor at 176 degrees Fahr. for three hours, after which it shall be immersed in hot water for forty-eight hours; another shall be placed in water at from 60 to 70 degrees Fahrenheit, and the third shall be left in moist air.

Samples of one-to-two mortar and of concrete shall be made and tested from time to time as directed by the superintendent. All cement shall be housed and kept dry till wanted in the work.

Storage rooms and rooms and apparatus for the tests shall be furnished by the contractor, and all tests shall be made entirely at his expense, and under the direction and to the satisfaction of the superintendent.

PORTLAND CEMENT CONCRETE.

The concrete shall be composed of clean, hard, broken limestone (or gravel with irregular surfaces) and cement mortar in volumes as hereinafter described. The sand shall be clean, sharp, Kansas River sand, washed *entirely* free from earth and loam. If obtainable, a mixture of coarse and fine sand shall be used. Approved mixing machines shall be used. These machines must be kept clean and no accumulations of old mortar shall be allowed to form in them. The ingredients shall be placed in the machine in a dry state and in the volumes specified and be thoroughly mixed, after which clean water shall be added and the mixing continued until the wet mixture is thorough and the mass uniform. No more water shall be used than the concrete will bear without quaking in ramming. The mixing must be done as rapidly as possible, and the batch deposited in the work without delay, and before the cement begins to set. Stone must be entirely free from earth and earthy surfaces. Thin splints or leaves of stone, easily broken with fingers, will not be allowed to go into the work. The quality of stone and the crushing must be acceptable to the superintendent.

The grades of concrete to be used are as follows (parts by volume):

For the arches: one part Portland cement, two parts sand, and four parts broken stone (hazelnut size, from one-half inch to one inch), except for the exposed faces and soffits of the arches, which shall have at least one inch in thickness of mortar composed of one part Portland cement and two parts sand.

For the piers, abutments, spandrel and wing walls: on the exposed surfaces for at least one inch thick one part Portland cement and two parts sand; for the next seven (7) inches one part Portland cement, two parts sand, and four parts broken stone of hazelnut size. For the remaining portions: one part Portland cement, four parts sand, and eight parts broken stone of size to pass through a three-inch ring; except such portions of the interior of the piers and abutments as are above the top of the cornice, or elevation 15.75 ft. above low water, which shall be composed of one part Portland cement, three parts sand, and six parts broken stone which will pass through a two and one-half inch ring.

No plastering of surfaces will be allowed nor any practice that will develop planes or surfaces of demarkation other than those hereinafter described. Immediately after the removal of any forms or centers, sand and cement shall be sifted on the surfaces and the surfaces rubbed hard with a float as may be directed by the superintendent.

During warm and dry weather and whenever the superintendent shall direct, all newly built concrete shall be kept well shaded from the sun and well sprinkled with water at the surface for several days or until well set.

There must be no definite plane or surface of demarkation between the facing and the concrete backing. The facing and the backing must be deposited in the same layer and well rammed in place at the same time.

In connecting old concrete with new, in the planes hereafter described, the old concrete shall be cleaned and roughened and soaked with water, and at the points of contact a mortar composed of one part cement and two parts sand shall be used and shall be laid in the same manner as specified for laying the facing.

NATURAL CEMENT CONCRETE.

The concrete around piles, to take the place of the earth excavated from the cofferdams, shall be composed of one part natural cement, equal to the best Fort Scott, Kas., cement, three parts sand, and six parts of broken stone of the size to pass through a three-inch ring. This concrete may be mixed by hand on platforms adjoining the foundations and shoveled directly into the coffer-dams. care being taken to deposit it in uniform layers of about six inches each and to carefully ram each layer.

PIERS, ABUTMENTS, AND SPANDRELS.

All piers, abutments, spandrels, and wing walls shall be built in timber forms. These forms shall be substantial and unyielding, of proper dimensions for the work intended and closely jointed, and all surfaces that come in contact with the concrete shall be smoothly dressed and well oiled with linseed oil to prevent the concrete from adhering to them. That portion next to the exposed faces of the work need not be oiled, but shall be covered with oiled paper, the same as that specified for the centers.

Molds, to form molding and panels, smoothly finished and well oiled with linseed oil, shall be properly placed in the forms so that the finished work will appear as shown on the plans. Extreme care must be used to place them in proper position before placing any concrete or mortar in them.

CONTINUOUS WORK.

The following divisions shall constitute sections for continuous work, viz.: each footing course of piers or abutments ; each pier or abutment from footing course to cornice ; each pier or abutment from cornice to springing line of arch ; each spandrel wall from keystone to pier or abutment ; each pier or abutment spandrel wall ; that portion of the piers or abutments above springing line of arch shall be considered part of the longitudinal sections of the arch previously described.

Each of the above sections shall be carried on continuously night and day if necessary ; that is, each layer shall be well rammed in place before the previously deposited layer shall have time to partially set.

Care shall be taken to make the joints (for expansion) in each spandrel wall over piers as indicated on the plans.

CONCRETE IN COFFER-DAMS.

The natural cement concrete in the coffer-dams shall extend from depths marked on plans to one foot below low water. Upon this concrete the footing courses of piers and abutments shall be founded.

The sheet piling of coffer-dams shall be cut off at least down to low-water mark, neatly and evenly, by the contractor before the completion of the work.

APPENDIX. 145

EXTRACTS FROM KATTE'S MASONRY SPECIFICATIONS.

By permission of WALTER KATTE, M. Am. Soc. C. E.

EXCAVATIONS will be classified under the following heads, viz.: earth, hardpan, loose rock, solid rock, and excavation in water.

EARTH will include clay, sand, gravel, loam, decomposed rock and slate, stones and boulders containing less than one cubic foot, and all other matters of an earthy nature, however compact, excepting only "hardpan," as described below.

HARDPAN will consist of tough, indurated clay or cemented gravel which, in the opinion of the Engineer, requires blasting for its removal.

LOOSE ROCK.—All boulders and detached masses of rock measuring over one (1') cubic foot in bulk, and less than one (1) cubic yard; also all slate, shale, soft friable sandstone and soapstone, and all other materials excepting rock, solid ledge, and those described above; also stratified rock in layers of not exceeding eight (8") inches in thickness, when separated by strata of clay, and which, in the judgment of the Engineer, may be removed without blasting, although blasting may occasionally be resorted to.

SOLID ROCK will include all rock found in ledges, or masses of more than one (1) cubic yard, which, in the judgment of the Engineer, may be best removed by blasting, with the exception of stratified rocks described under the head of Loose Rock. In rock excavations the "bottom" must in all cases be taken down truly to sub-grade; and when so ordered by the Engineer ditches must be formed at the foot of the slope.

The contract price for excavation will apply to pits required for foundations of masonry when water is not encountered, and the price for

EXCAVATION IN WATER will only apply to foundation pits under water and deepening of channels in running water; it must cover all classes of material, and include drainage, bailing, pumping, and all materials and labor connected with such excavations; also the necessary dressing of the rock.

CEMENT must be of the best quality of freshly burned and ground hydraulic cement, and be equal in quality to the best brands of   Cement. It will be subject to test made by the Engineer or his appointed inspector, and must stand a proof tensile test of fifty (50) pounds per square inch of sectional area on specimens allowed a set of thirty (30) minutes in air and twenty-four (24) hours under water.

MORTAR will in all cases be made of one part in bulk of the best hydraulic cement to two parts in bulk of clean, sharp sand, well and thoroughly mixed together in a clean box of boards, before the addition of the water, and must be used immediately after being mixed. No mortar left over night will, under any pretext, be allowed to be used. The sand and cement used will at all times be subject to inspection, test, acceptance, or rejection by the Engineer.

CONCRETE.—Concrete shall be composed of fragments of hard, sound, and acceptable stone, broken to a size that will pass through a two (2") inch ring in any direction, thoroughly clean and free from mud, dust, dirt, or any earthy admixture whatever; mixed in the proportion of two (2) parts in bulk of the broken stone to one (1) part of fresh-made cement mortar of the quality herein described; and is to be quickly laid in sections and in layers not exceeding nine (9) inches in thickness, and to be thoroughly rammed until the mortar flushes to the surface; it shall be allowed at least twelve (12) hours to "set" before any work is laid on it.

FOUNDATIONS.

GENERAL DESCRIPTION. — Foundations for masonry shall be excavated to such depths as may be necessary to secure a solid bearing for the masonry, of which the Engineer shall be the judge. The materials excavated will be classified and paid for, as provided for in these specifications, under the general head of Excavations; and in case of foundations in rock, the rock must be excavated to such depth and in such form as may be required by the Engineer, and must be dressed level to receive the foundation course.

When a safe and solid foundation for masonry cannot be found at a reasonable depth (to be judged of by the Engineer), there will be prepared by the contractor such artificial foundations as the Engineer may direct. All materials taken from the excavations for foundations, if of proper quality, shall be deposited in the contiguous embankment; but any material unfit for such purpose shall be deposited outside the roadway, or in such place as the Engineer shall direct, and so that it shall not interfere with any drain or watercourse.

TIMBER. — Timber foundations when required shall be such as the Engineer may by drawings or otherwise prescribe, and will be paid for by the one thousand feet, board measure. The price, covering cost of material, framing and putting in place, and all wrought- and cast-iron work ordered by the Engineer, will be paid for per pound, the price including cost of material, manufacture, and placing in the work.

PILING. — All timber used in foundations or foundation piling shall be of young, sound, and thrifty white oak, yellow pine, or other timber equally good for the purpose, acceptable to the Engineer. Piles must be at least eight (8″) inches in diameter at the small end and twelve (12″) inches in diameter at the butt when sawn off; they must be perfectly straight and be trimmed close, and have the bark stripped off before they are driven. They must be driven into hard bottom until they do not move more than one-half inch under the blow of a hammer weighing two thousand (2000) pounds, falling twenty-five (25′) feet at the last blow. They must be driven vertically and at the regular distances apart from centers, transversely and longitudinally, as required by the plans or directions of the Engineer; they must be cut off squarely at the butt and be well sharpened to a point, and when necessary, in the opinion of the Engineer, shall be shod with iron and the heads bound with iron hoops, of such dimensions as he may direct, which will be paid for the same as other iron work used in foundations.

The necessary length of piles shall be ascertained by driving test piles in different parts of the localities in which they are to be used; and in case a pile shall not prove long enough to reach "hard bottom" it shall be sawed off square, and a hole two (2″) inches in diameter be bored into its head twelve (12″) inches deep; into this hole a circular white-oak trenail twenty-three (23″) inches in length shall be well driven, and another pile similarly squared and bored, and of as large a diameter at the small end as can be procured, shall be placed upon the lower pile, brought to its proper position, and driven as before directed. All piles, when thus driven to the required depth, are to be cut off truly square and horizontal at the proper height given by the Engineer, and only the actual number of lineal feet of the piles left for use in the foundations after being sawn off will be paid for.

COFFER-DAMS. — Where coffer-dams are, in the opinion of the Engineer, required for foundations the prices provided in the contract for timber, piles, and iron in foundations will be allowed for the material and work on same, which is understood as covering all risks from high water or otherwise, draining, bailing, pumping, and all materials connected with the coffer-dams. Sheet piling will be classed as plank in foundations, and will be paid for per one thousand (1000′) feet board measure if left in the ground.

TIMBER.

All timber must be sound, straight-grained, and free from sap, loose or rotten knots, wind shakes, or any other defect that would impair its strength or durability; it must be sawed (or hewed) perfectly straight and to exact dimensions, with full corners and square edges; all framing must be done in a thoroughly workmanlike manner, and both material and workmanship will be subject to the inspection and acceptance of the Engineer.

SPECIFICATIONS FOR STEEL COFFER-DAM.

DESIGN.—The shell shall be made of elliptical shape for ordinary piers and circular for pivot piers. It shall be made not less than four feet larger than footing of pier in plan, to allow for variation in position during sinking.

The plates used shall be as large as can be handled with ease in the shop, during shipment, and during erection.

The splices may be either lap or butt joints, provided a good tight job will result, and the rivets must be spaced according to boilermaker's rules.

The joint may be made tight by calking or by the use of a calking strip, but in either event the result must be guaranteed.

The shell must be stiffened by horizontal stiffening angles, girders, or trussing, to resist deformation during the placing and to resist both the quiescent and a maximum unbalanced earth or water pressure, or a wind pressure.

The bottom plates shall be re-enforced with narrow plates inside and outside, to form a wedge-shaped cutting edge; and when there is rock or hard bottom the plates shall be cut to conform to its contour as nearly as possible.

The top shall be properly stiffened, and if necessary provided with connection holes for additional sections.

The factor for safety shall in no case be less than four, and in case the shell will be subject to shock, not less than six.

No metal of a less thickness than 1/4 inch shall be used for temporary work, nor less than 3/8 inch for permanent work in fresh water or 1/2 inch in salt water.

MATERIAL.—The entire shell shall be constructed of the grade of steel known as soft medium, except rivets, which shall be of bridge quality of iron.

The steel may be made either by the Bessemer or open-hearth process, and the phosphorus shall never exceed 0.08 per cent.

Soft medium steel shall have an ultimate strength of from 55,000 to 65,000 pounds per square inch, as determined from standard test pieces ; an elastic limit of not less than one-half the ultimate strength ; an elongation of not less than 25 per cent in 8 inches , and a reduction of area at fracture of not less than 50 per cent.

Samples to bend cold 180 degrees to a diameter equal to the thickness of the sample, without crack or flaw on the outside of the bent portion.

ERECTION.—The erection must be done in a first-class manner, and all rivets must have full heads. The shell shall be placed in position within one-half the distance allowed for error in the design of the coffer-dam. Only a reasonable variation will be allowed for difference in level.

PAINTING.—All the metal work shall be thoroughly cleaned of rust or scale at the shops and coated thoroughly with hot asphaltum.

Before erection, in the field, it shall be given a second coating of hot asphaltum.

SEALING.—When in position on the bottom, if the coffer-dam has not been sunk through impervious strata, it shall be sealed by concreting around the circumference inside with concrete passed through a tube.

REMOVAL.—Should the coffer-dam not form a part of the permanent foundation it shall be taken apart, at the joints designed for the purpose, and carefully removed in such a manner as not to injure the foundation, and so as to be used again if required.

HEALD & SISCO STANDARD IRON HORIZONTAL CENTRIFUGAL PUMPS.

No.	Capacity in Gallons per Minute.	Horse-power required for Each Foot of Lift. Minimum Quantity.	Diameter and Face of Pulley in Inches.	Floor Space required, in Inches.	Shipping Weight. Pounds.	Price of Pump, Oilers and Wrench.	Price of Pump and Primer.	No.
1½	50 to 70	.024	6 × 6	17 × 30	168	$45	$55	1½
1¾	75 to 100	.037	7 × 8	21 × 33	232	60	70	1¾
2	110 to 150	.054	8 × 8	23 × 37	306	75	90	2
2½	175 to 250	.086	8 × 8	24 × 38	348	90	105	2½
3	250 to 350	.124	8 × 8	25 × 39	400	110	130	3
4	450 to 600	.223	10 × 10	30 × 40	545	130	155	4
5	750 to 900	.372	15 × 12	34 × 54	826	165	195	5
6	1000 to 1400	.496	15 × 12	37 × 55	965	200	240	6
8	1700 to 2200	.844	20 × 12	45 × 63	1500	310	375	8
10	2200 to 3000	1.093	24 × 12	51 × 71	2170	395	470	10
12	3000 to 4000	1.49	30 × 14	62 × 75	3050	500	...	12
15	4800 to 6000	2.38	40 × 15	77 × 80	7100	850	...	15
*15	4800 to 6000	2.38	30 × 15	60 × 68	3150	710	...	15
18	7500 to 10000	3.73	40 × 15	93 × 103	9000	1300	...	18
*18	7500 to 10000	3.73	30 × 16	62 × 70	3500	1150	...	18
22	12000 to 14000	5.96	48 × 20	126 × 130	12000	...	...	22

* Refers to low-lift pump.

The number of pump is also diameter of discharge opening in inches. Where more than 25 feet of discharge pipe is attached to pump, use one or two sizes larger than pump discharge.

For No. 12 and larger sizes a foot valve or flap valve and ejector for priming is recommended.

LIST OF HEALD & SISCO HYDRAULIC DREDGING AND SAND PUMPS.

Number of Pump.	Diameter Suction and Discharge Openings. Inches.	Cubic Yards of Material they will Raise per Hour.	Horse-power recommended for 10-Foot Lift.	Diameter and Face of Pulley.	Floor Space Required. Inches.	Shipping Weight. Pounds.	Will Pass Solids, Diameter. Inches	Price of Pump Complete, with Suction and Discharge Elbows, Flap Valve and Ejector.	Number of Pump.
4	4	30 to 50	6	12 × 12	40 × 31	800	2	$210	4
6	6	60 to 80	12	20 × 12	68 × 40	1700	4½	300	6
8	8	125 to 150	22	24 × 14	72 × 18	3400	6	475	8
10	10	200 to 300	35	30 × 15	94 × 54	4200	8	600	10
12	12	300 to 375	45	36 × 20	114 × 66	9000	10	850	12
15	15	400 to 500	75	42 × 24	154 × 78	12000	10	1450	15
18	18	500 to 700	125	48 × 30	160 × 80	13500	10	1900	18
20	20	....	...	....	....	.....		....	20
22	22	....	...	....	....	.....		....	22

NUMBER OF REVOLUTIONS AT WHICH PUMPS SHOULD RUN TO RAISE WATER TO DIFFERENT HEIGHTS.

No.	5 Feet.	10 Feet.	15 Feet.	20 Feet.	25 Feet.	30 Feet.	35 Feet.	40 Feet.
1½	428	604	739	854	955	1045	1131	1208
1¾	348	491	601	695	777	850	920	982
2	302	426	522	603	674	737	798	852
2½	302	426	522	603	674	737	798	852
3	302	426	522	603	674	737	798	852
4	285	402	493	569	637	697	754	805
5	256	362	443	512	572	626	678	724
6	214	302	368	427	478	523	566	604
8	183	259	317	366	409	448	485	517
10	168	238	291	336	376	411	445	475
12	133	188	230	266	298	326	352	376
15	105	148	181	209	234	256	277	295
*15	151	213	261	301	337	369	399	426
18	105	148	181	209	234	256	277	295
*18	151	213	261	301	337	369	399	426

* Refers to low-lift pumps.

Above table gives *correct* speed of pumps as employed under usual conditions of pumping. If water must be forced through a number of bends and elbows, or a great length of piping, the above speed must be somewhat increased.

Use large pipes and easy bends wherever practicable, as they save power.

TABLE OF SIZES, LIDGERWOOD SINGLE-CYLINDER, SINGLE-DRUM HOISTING-ENGINES.

Horse-power Usually Rated.	Dimensions of Cylinder.		Weight Hoisted Single Rope Usual Speed. Lbs.	Suitable Weight of Pile-driving Hammer for Quick Work. Lbs.	Dimensions of Hoisting-drum.			Dimensions of Bed-plate.		Dimensions of Boiler			Estimated Shipping Weight Compl.etc. Lbs.
	Diameter. Inches.	Stroke. Inches.			Diam Body between Flanges. Inches.	Length Body between Flanges. Inches.	Diameter Flanges. Inches.	Width. Inches.	Length. Inches.	Diameter Shell. Inches.	Height Shell. Inches.	Number of 2-inch Tubes.	
4	5	8	1200	1000	10	20	22	38	60	28	63	40	3550
6	6¼	8	1500	1250	10	20	22	38	60	28	69	40	3950
8	6¼	10	1750	1500	12	20	24	41	73	30	72	44	4850
10	7	10	2500	1800	12	20	24	41	73	32	75	48	5050
11	7	10	2500	2000	14	22	26	45	73	34	78	52	5350
12½	8¼	10	4000	2500	14	23	29	47	73	36	75	57	6550
15	8¼	10	4000	2800	14	23	29	47	73	36	81	57	6750
20	8½	12	6000	4000	16	26	33	54	84	40	84	80	8500
25	10	12	8000	5000	16	26	33	54	84	42	90	88	9500

TABLE OF SIZES, LIDGERWOOD DOUBLE-CYLINDER, DOUBLE-DRUM HOISTING-ENGINES.

Horse-power Usually Rated.	Dimensions of Cylinders.		Dimensions of Hoisting-drums		Weight Hoisted Single Rope. Average Speed.	Suitable Weight of Pile-driving Hammer for Quick Work.	Dimensions of Boiler.			Dimensions of Bed-plate.		Estimated Shipping Weight with Boiler Complete.
	Diam. Inches.	Stroke. Inches.	Diam. Inches.	Length. Inches.			Diam. Inches.	Height. Inches.	Number of 2-inch Tubes.	Width. Inches.	Length Inches.	
8	5	8	12	22	2000	1500	32	75	48	47	80	6500
12	6¼	8	14	22	3000	2000	36	75	57	50	86	8000
16	6¼	10	14	26	4000	2800	38	81	68	54	89	9000
20	7	10	14	26	5000	3500	40	84	80	54	89	9550
30	8¼	10	14	27	8000	5000	42	90	88	57	94	11400
40	8½	12	16	32	10000	8000	50	102	124	70	117	21000
50	10	12	16	32	12000	10000	53	102	150	70	117	22000

INDEX.

Aa river, Russia, 83
Ancient methods of—
  Founding, 1, 3, 4, 5
  Pile driving, 5, 40, 41
  Pumping, 92, 93
  Sheet piling, 6, 7
Anchoring, coffer-dam, 30, 32, 33, 87
Approval, War Dep't, 120
Arkansas river, 20, 70
Architectural design, piers, 125, 127, 131
Arch bridge—
  Center, 141
  Hutcheson, Scotland, 6
  Largest, 3
  Melan, Topeka, 78, 119, 141, 142
  Roman, 1
  Shuster, Persia, 1
  Topeka, Kansas, 78, 119, 141, 142
  Trezzo, 1
Asphalt for leaks, 65
Bamboo casings, 4
Bank protection—
  Japanese, 4
  Mississippi, 4
Bascule pump, 92
Batter of piers, 125
Bear river, Canada, 83
Bearing—
  Piles, 106, 146
  Power of piles, 49
Blasting, 33, 73, 89
Boiler riveting, 80, 82, 147
Bolts, 7, 57, 139
  See Drift-bolts
Borings—
  Auger, Pierce, 121
  Casing for, 121
  Clamp for driving tube, 122
  Core removal, 122
  Cutting shoe for pipe, 122
  Drills for, 121
  Driving pipe for, 122
  Extensive, 121
  Hand drills for, 121
  Jars for, 121
  Maul for driving pipe, 122

Borings—*Continued.*
  Obstructions to, 121
  Pebble-tongs, 124
  Pump, sand, 121, 122
  Removal of core, 122
  Rope knives, 121
  Rope spears, 121
  Sand pump, 121, 122
  Screw, adjusting, 121
  Spears, rope, 121
  Temper screw, 121
  Test, 121
  Tongs, pebble, 124
  Tripod for, 121
  Well-driller, 121
Bottom—
  Clay, 38, 57, 60, 73, 77
  Drift in sand, 18
  Gravel, 6, 16, 20, 57, 60, 64, 66, 74, 106, 133
  Hard clay, 10, 12, 77, 106
  Mud, 17, 38, 60, 62, 77, 106
  Open (porous), 16, 28
  Overlaid. *See* Rock
  Porous, 16, 28, 141
  Quicksand, 67
  Rock, 1, 20, 24, 25, 26, 28, 29, 30, 32, 33, 36, 38, 39, 57, 62, 66, 72, 74, 77, 86, 106
  Rock, overlaid, 16, 20, 24, 25, 29, 33, 36, 38, 57, 66, 72, 74, 77, 106
  Sand, 62, 64, 68, 77, 78, 106
  Sand and drift, 18
  Sand and shells, 77
  Shale, 60
  Shells in sand, 77
  Silt, 33
  Soapstone, 20
  Soundings, 86, 120, 121
  Uneven, 17, 32, 33, 36, 62, 76, 106
Boxing in leaks, 32, 39, 60
Box pump, 93
Bras d'Or river, Canada, 82
Brace rods, 30, 38, 67, 134
Braces, 7, 30, 33, 38, 39, 56, 64, 67, 76 80, 83, 84, 135
Bridge location, 120
Bridge cost, economic, 124, 125

151

152 INDEX.

Bridges referred to—
  Ann Arbor, 59
  Arnprior, 18
  Arthur Kill, 38, 60
  Baltimore, N. Ave., 118
  Blackfriars, London, 4
  Boucicault France, 109
  Buda-Pesth, 8, 28, 50, 53
  Cæsar's, over Rhine, 5, 40
  Charlestown, Boston, 53, 63
  Chattanooga, Walnut St., 66, 99, 107
  Coteau, 39
  Cumberland, Md., 67
  Fair Haven, 48
  Forth, Scotland, 86, 99, 107
  Fort Madison, Ia., 20
  Gadsden, Ala., 74
  Harvard, Boston, 110
  Harlem ship-canal, 36
  Harper's Ferry, 63
  Hawkesbury, Australia, 80
  Hutcheson, Glasgow, 6, 50, 57
  Little Rock, Ark., 70, 111
  Melbourne, Australia, 33
  Momence, Illinois, 62
  Neuilly, France, 92
  Omaha, Nebraska, 127
  Orleans, France, 40, 92, 93
  Philadelphia, Walnut St., 24
  Phila. & Reading R. R., 74
  Putney, England, 77
  Red River, 111
  Riga-Orel, Russia, 83
  Rochester, Court st., 117
  Saumur, France, 41
  Shuster, Persia, 1
  Topeka, Kansas, 78, 119
  Trajan's, 40
  Tulsa, 20
  Victoria, Canada, 83
  Westminster, London, 4
Bucket-wheel pump, 92
Bucket for concrete, 36, 39, 110
Bull-wheel pile driver, 41
Bulkhead, 30
Cableway—
  Capacity of, 117
  Carriage for, 117
  Span of, 117, 118
  Use of, 117
Caisson—
  Open, 3, 4, 13
  Pneumatic, 5
  Water-tight, 74
Calking—
  Cylinders, 82
  Joints, 20, 36, 147
Canal—
  Chicago drainage, 117
  Harlem ship, 36
  Illinois & Miss., 113
  Keokuk, Ia., 30
  N. Y. State, 45, 50

Candle wick for leaks, 20
Cane stalks for leaks, 28
Canvas sheet, 30, 31, 32, 33, 35
Capacity of—
  Cableway, 117
  Dredge, 101, 102, 103, 135, 148
  Pumps, 77, 78, 93, 94, 99, 101, 135, 148
Carriage for cableway, 117
Casing—
  Bamboo, 4
  Boring, 121
  Timber, 5
Cement (*see* Concrete)—
  Defective, 114, 136
  Laitance of, 109
  Mortar, 136, 143, 145
  Natural, 111, 136, 143, 145
  Portland, 113, 115, 116, 142
  Quality of, 114, 136, 142, 145
  Specifications, 136, 142, 145
  Tests of, 115, 136, 142, 145
Center for arch, 141
Centrifugal pump, *see* Pump
Chamber, width of, 56
Chamber, *see* puddle
Changes, 139
Channel—
  Establishing new, 3
  Fixed place for, 120
  Requirements of Gov't, 120
Chapelet pump, 92, 93
Charles river, Boston, 63, 110
Circular—
  Coffer-dam, *see* Coffer-dam
  Pier of granite, 86
  Shell for pier, *see* cylinder
Clam-shell dredge, 102
Clamp for—
  Coffer-dam, 38
  Pile-driver, 43
  Pipe, 122
  Sheet piles, 57
Classification of excavation, 135, 145
Clay, *see* Bottom
Clay puddle, *see* Puddle
Clearance in coffer-dam, 74, 147
Clyde river, Scotland, 6
Coffer-dam—
  Anchoring, 30, 32, 33, 37
  Calculation of, 54, 56, 84
  Canvas and plank, 30, 32, 33, 35
  Circular, 20, 24, 87, 147
  Clearance in, 74, 147
  Cost of, 61, 76, 77
  Crib type, 14, 15, 17, 18, 20, 26, 32, 33, 36, 38, 39, 72
  Damaged, 24, 32, 35, 60, 66, 76
  Decking for, 134
  Definition of, 13
  Deposit in, 134
  Earth bank type, 5, 13, 14, 26
  Economy in, 26
  Erecting steel, 147

Coffer-dam—*Continued.*
  Failure of, 16, 24, 28, 30, 32, 60
  Floating type, 74
  Frame for, 30, 35, 59, 66, 67, 75
  Grillage type, 20, 24
  Half-tide, 89
  Largest recorded, 8
  Location of, 86, 87
  Metal, type of, 83, 86, 147
  Movable, 20, 33, 74, 75, 76
  Moving, time required, 76
  Origin of, 3, 5
  Pivot pier, 20, 36, 38, 39, 147
  Polygonal type, 36, 38, 39
  Price for, 139
  Protection, 26, 33
  Puddle, pressure on, 54
  Puddle for, *see* Puddle
  Removal of, 120, 33, 74, 134, 147
  Removing piers by, 72, 74
  Robinson, circular, 24
  Sheet pile type, 6, 8, 9, 24, 26, 59, 60, 61, 62, 63, 64, 65, 134
  Sinking with stone, 20, 76
  Specifications for, 6, 134, 141, 146, 147
  Splicing in height, 67, 147
  Tarpaulin and plank, 30, 32, 33, 35
  Tidewater type, 8, 53, 63, 86
  Wakefield piling, 14, 20, 53, 61, 62, 63, 64, 141
  Water pressure on, 54, 56
  Width of chambers, 56
Compartments, water-tight, 76
Completion of work, 139
Composition of concrete, *see* Concrete
Compound sheet piles, 52, 53, 61, 66
Concrete—
  Ancient use of, 2, 4
  Bucket for depositing, 36, 39, 110
  Composition of, 107, 110, 111, 116, 138, 143, 145
  Cost of, 117
  Depositing, rate of, 115
  Facing for, 72, 111, 114, 143
  Forms for, 111, 113, 115, 141, 142, 144
  Foundation of, 18, 24, 36, 60, 61, 63, 67, 72, 107, 108, 116, 117, 138, 143, 144, 146
  Laitance, 109
  Laying, rules for, 113, 138, 143, 145
  Layers, proper thickness, 109, 113, 138, 143, 144, 145
  Leveling course, 107, 117
  Louisville cement used, 111
  Monolithic construction, 72, 111, 112, 113, 143
  Oil paper on forms, 142
  Piers of, 72, 111, 143, 144
  Pier filling, 80, 82, 143
  Portland cement used, 72, 113, 115, 116, 143
  Proportions of, 67, 72, 107, 110, 111, 116, 138, 143, 145

Concrete—*Continued.*
  Puddle of, 38, 78, 87
  Rate of deposit, 115, 143, 144, 145
  Rules for laying, 113, 138, 143, 144, 145
  Sacked for placing, 107
  Sand for, 111, 115, 136, 143, 145
  Setting time, 110, 114, 143, 144, 145
  Stone for, 111, 115, 143, 145
  Tube for depositing, 108, 110
  Under water, 36, 39, 107, 108, 109, 110, 142
  Water, amount for, 113, 114, 143
  Wells in, 114
Coosa river, Ala., 74, 110, 117
Coping, 125, 127, 129, 137
Corbel course, 127, 129
Core removal, test, 122
Corporation, applicant for bridge, 120
Cost of—
  Bridge, least, 124
  Coffer-dam, 61, 76, 77
  Concrete, 117
  Dredgers, 103
  Dredging, 103, 148
  Driving piles, 48
  Formula for, spans, 124
              bridge, 124
              piers, 125
  Hoist-engines, 44, 45
  Piers, 125
  Pumps, 93, 148
  Removing pier, 74
  Spans, 124
Crevices in rock, *see* Rock
Cribs, 26, 33, 111
Crib anchor, 33, 134, 135
Crib coffer-dam, *see* Coffer-dam
Current, strength of, 120
Cutting edges, 80, 122, 147
Cutting shoe, 122
Cutwaters, 10, 18, 26, 66, 129, 130, 131
Cylinder—
  Bracing for, 83, 84
  Calking for, 82
  Guide for, 81
  Pier, 33, 80, 82, 83, 87
  Piles for, 83
  Thickness of, 82, 83, 84
Damaged—
  Coffer-dam, 24, 32, 35, 60, 66, 76
  Piles, 10
Danube river, 9, 40
Decking for coffer-dam, 134
Defects in—
  Cement, 114, 136
  Piles, 10
  Stone, 137
Deposit in coffer-dam, 134
Depositing concrete under water, *see* Concrete
Derricks, 102, 117
Derrick, *see* Pile-driver
Design of piers, *see* Piers
Direct connected pump, 97

INDEX.

Disposal of excavation, 26, 60, 71
Diver employed, 31, 36, 76, 87
Docks, Victoria, B. C., 77
Dowels for stone, 137
Drawings to show—
  Bridge location, 120
  Bridge plan, 120
Dredger—
  Capacity of, 101, 102, 103, 135, 148
  Clam-shell, 102
  Claw, 72, 80, 102
  Cost of, 103, 148
  Derrick for, 102
  Dipper, 103
  Edward's Cataract pump, 101
  Elevator type, 102
  Engine for, 78, 148
  Furnished, 134
  Grapple, 102
  Lancaster, 102
  Osgood, 103
  Pump, 78, 99, 101, 148
  Sand-digger, 102
  Scraper, 13
  Seagoing, 101
  Spoon, 13
Dredging, 10, 15, 80
  Amount necessary, 16
  Cost of, 102, 103, 105
  Pumps for, 78, 99, 101, 148
  Rock, 72
  Soft bottom, 106
  Wells for, 80
Drift in sand, 18
Drift bolts, 33, 36 38, 39, 57, 60, 71
Drilling, test, see Boring
Driver, see Pile-driver
Durability of piles, 40, 72
Earth bank coffer-dam, see Coffer-dam
Economic—
  Bridge cost, 124, 125
  Coffer-dam construction, 26
  Pier spacing, 120, 124, 125
Eddies, 129, 130, 131
Efficiency of pumps, 96, 148, 149
Ejector for priming, 101, 148
Electricity—
  Blasting by, 73
  Hoisting by, 98, 117
  Pumping by, 97
Engine, see Hoist
Erection, steel coffer-dam, 147
Estimates, 139, 146
European pier design, 127
Examination for bridge site, 120
Excavation (see Dredging) — 7, 10, 60, 61
  Classification, 135, 145
  Disposal of, 26, 60, 71, 135
  Measurement, 139
  Rock, 72, 86, 145
  Scraper for, 13
  Spoon for, 14

Experiments—
  Piers, form of, 129, 130, 131
  Puddle, 12
  Timber, wet, 33
Exterior puddle, see Puddle
Facing for concrete, 72, 111, 114, 143
Failure of—
  Coffer-dam, see Coffer-dam
  Contractor to prosecute work, 139
  Puddle, 28
Filling, see Puddle
Floating coffer-dam, 74
Footing course, 4, 117, 127
Form of—
  Foundation, 5, 124
  Piers, 129, 130, 131
  Sheet piles, 50, 51, 52, 53
Forms for concrete, 111, 113, 115, 141, 142, 144
Formula for—
  Cost of hoists, 45
  Cylinder thickness, 84
  Economic span, 124
  Pile loads, 49
  Sheet pile thickness, 54, 56
  Struts, 56
Fort Monroe, Va., 64
Forth Bridge piers, 86
Foundations—
  Ancient, 1, 3, 4, 5, 40
  Care, increased, 1
  Changes in, 136
  Character of, 106, 117, 124, 146
  Coffer-dam, origin, 3, 5
  Concrete, see Concrete
  Crib, early type, 5
  Difficult, very, 120, 124
  Doubt of obtaining, 125, 146
  Encaissement, 3
  Footing course, 4, 117, 127
  Form of, 5, 124, 146
  Grillage, 20, 24, 60, 67, 71, 72, 106
  Open caissons, 3, 4, 13
  Origin, sub-aqueous, 3
  Piles and concrete, 3, 106, 110
  Piles, bearing, 106, 146
  Piles under cylinders, 83
  Pneumatic caisson, 5
  Risk of, 125
  Rock bottom, see Bottom
  Roman, 1
  Steel shells, 33, 80, 82, 83, 87
  Sub aqueous, 3, 77
  Tropical, 1
Frame for coffer-dam see Coffer-dam
Freshet, damage by, 66, 76
Friction lever, 44
Frost on mortar, 136
Gate valve for priming, 101
Girders, circular, 89, 147
Government—
  Approval, 120
  Requirements, 120

INDEX.

Grades, 135
Gravel bed, *see* Bottom
Gravel in puddle, 12, 15, 26, 76, *see* Puddle
Grease for leaks, 30, 36
Grillage (*see* Foundation)—
  Removal, 74
Guide—
  Cylinder pier, 81
  Piles, 7, 49, 57, 71
  Pile-driver, 43
  Pipe for drilling, 121
Guide for piles, 43, 57
Gunny sacks, 26, 28, 33, 36, 64, 76
Half-tide coffer-dam, 89
Hammer, *see* Pile-driver
Hand—
  Derrick, 40, 41, 43
  Pump, 93
Harlem ship-canal bridge, 36
Hoist engine for—
  Derrick, 117, 149, 150
  Pile-driver, 44, 45, 68, 78, 149, 150
  Scraper, 13
Hoist—
  Cost of, 44, 45
  Electric, 98, 117
  Weight of, 149, 150
Huron river, Mich., 59
Hutcheson bridge, 6
Ice protection, 9, 10, 26, 129, 130, 131
Illinois & Miss. canal, 113
Illinois river, 62
Inspection, 139
Iron coffer-dam, 83, 86
Iron sheet piles, 91
Jet for pile-driver, 70
Kanawah river, W. Va., 14, 117
Kankakee river, 62
Karun river, Persia, 1
Key piles, 57
Knives for drill-rope, 121
Laitance of cement, 109
Lansdell siphon, 93
Largest—
  Arch, 3
  Coffer-dam, 8
Layers of concrete, 109, 113, 138, 143, 144, 145
Laying concrete, rules for, 113, 138, 143, 145
Leaks cured by—
  Asphalt, 65
  Boxing in, 32, 39, 60
  Calking joints, 20, 36
  Candle wicking, 20, 39
  Cane-stalks crushed, 28
  Canvas funnel, 32
  Clay cylinders, 29
  Clay dams, 89
  Concrete in sacks, 66
  Exterior puddle, 28, 73
  Gravel and clay, 26
  Grease, 30, 36

Leaks cured by—*Continued.*
  Grouting, 89
  Hot asphalt, 65
  Manure, 28, 32
  Packing puddle, 12, 28, 77
  Puddle, exterior, 28, 78
  Puddling rock seams, 25
  Repuddling, 28
  Round braces, 30, 38
  Sacks, 26, 36, 66, 76
  Sheet piles, 12, 28
  Stiff grease, 30, 36
  Stock ramming, 12, 23, 77
  Straw and gravel, 20, 28
  Tarpaulin, 30, 32, 33, 35, 36, 65
  Washers on rods, 30
  Water-head, 32, 39
Ledges, *see* Rock
Length, economic span, 124, 125
Leveling course concrete, 107, 117
Lift of pumps, 96, 100, 148, 149
Little Bras d'Or river, Canada, 82
Location of—
  Bridge, 120
  Coffer-dam, 86, 87
  Piers, 120
  Pumps, 100
Manure for leaks, 28, 32
Map of bridge location, 120
Maslin pump, 96
Masonry—
  Ashlar, 136
  Footing courses, 74, 117
  Laying, 137
  Marking stones, 74
  Measurement, 139
  Removing, 73, 74
  Rubble, 86, 90
  Specifications for, 136, 137, 138
Mass. sewerage system, 60, 61
Mattresses, 4
Maul for driving pipe, 122
Maul as pile-driver, 40
Measurement of work, 139
Melan Arch, Topeka, 78
Metal coffer-dam, 35, 83, 86, 147
Missouri river, 127
Mississippi river, 20, 32, 67
Monolithic concrete, 72, 111, 112, 113, 143
Morison's pier design, 125
Mortar, cement, 136, 143, 145
Movable coffer-dam, *see* Coffer-dam
Mud, *see* Bottom
Nasmyth hammer, 14, 45, 46, 47, 48, 49
Natural cement, 111, 136, 143, 145
Navigable rivers, *see* River
Navigation, needs of, 120
  Interference with, 135
New York State canals, 45, 50
Nippers, pile-driver, 43
Obstruction—
  To boring, 121
  By piers, 125

Ohio river, 15, 133
Oil paper for concreting, 142
Origin of—
  Coffer-dam, 3, 5
  Foundations, 3
Packing puddle, 12, 28, 77
Painting, 65, 147
Parnitz river, Germany, 72
Payment, manner of, 134
Pierce boring auger, 121
Piers—
  Architectural design, 125, 127, 131
  Batter of, 125
  Bracing for cylinders, 83
  Concrete, 72, 111, 143, 144
  Coping for, 125, 127, 129
  Cost of, for economy, 125
  Cylinder, 33, 80, 82, 83
  Design of, 125, 127, 129, 130, 131
  Economic spacing, 120
  European design, 127
  Experiments on form, 129, 130, 131
  Filling of concrete, 80, 82, 143
  Forth bridge, 86
  Hutcheson 6
  Location of, 120, 125
  Morison's design, 125
  Obstruction of, 125, 129, 130, 131
  Pivot, 82
  Relation to spans, 1, 124
  Removal, 72, 74
  Starlings of, 125, 129, 130, 131
  Thickness of tubular, 84
  Tubular steel, 33, 80, 82, 83
Pile-driver—
  Ancient, 5, 40, 41
  Beetle for, 40
  Bull-wheel for, 41
  Clamps, 43
  Cost of outfit, 43, 44
  Cram-Nasmyth, 49
  Derrick, 41, 42, 43, 45
  Engine, 44, 45, 78, 149, 150
  Friction lever, 44
  Guides, 43
  Hammer, 41, 43, 44, 68, 77, 78
  Hand derrick, 40, 41, 43
  Hoist engine for, 44, 45, 68, 78, 149, 150
  Horse-power, 41, 43
  Lidgerwood, 43
  Maul, 40
  Nasmyth, 14, 45, 46, 47, 48, 49
  Nippers, 43
  Rock drill used, 14
  Scow for, 43, 44, 45
  Sheet pile, 14, 41
  Sledge, 40
  Tongs, 43
  Warrington-Nasmyth, 46, 47
  Water jet, 70
  Windlass, 40
Piles—
  Ancient use, 3, 4, 5, 40

Piles—*Continued*.
  Bearing, 106, 146
  Bearing power, 49, 146
  Blasting out, 73
  Cost driving, 48
  Clamps, 43
  Damaged, 10
  Durability of, 40, 72
  Guide 7, 57, 49, 71
  Guiding, 43
  Key, 57
  Payment for, 139, 146
  Pointing, *see* Pointing
  Protection, 134
  Pulling, 50, 73
  Pulling lever, 50
  Pulling scow, 50
  Rings for, 70
  Sawing off, 50, 83, 146
  Sheet piles, ancient, 4
    See sheet piles
  Shoes, 50, 53
  Specifications for, 135, 141, 146
  Splitting of, 43
  Temporary, 135
  Under cylinders, 83
Pivot pier—
  Cylinders for, 82
  Coffer-dam for, 20, 36, 38, 39, 147
Pointing piles, 24, 50, 51, 53, 57, 60
Pointing with mortar, 136
Porous bottom, 16, 28, 141
Portland cement, 72, 113, 115, 116, 142
Potomac river, 63
Primers for pumps, 100, 101, 148
Pressure of—
  Puddle, 54
  Water, *see* Water
Proportions for concrete, 67, 72, 107, 110, 111, 116, 138, 143, 145
Protection of—
  Bank, 4
  Coffer dam, 26, 33, 134
  From ice, *see* Ice
Puddle—
  Blue clay, 72
  Chamber, 7, 10, 33, 36, 38, 56, 66, 72, 74, 77
  Clay, 7, 17, 26, 33, 66, 88
  Clay and gravel, 12, 15, 26, 76, 134
  Concrete, 39, 78, 87
  Experiments, 12
  Exterior, 17, 20, 25, 26, 28, 33, 63, 66, 71, 76, 77
  Failure of, 28
  Pressure of, 54
  Sacked, 26, 28, 33, 64
Pulling piles, 50, 73
Pulling test tubing, 124
Pulsometer, 30, 32, 90, 94, 95
Pumping, 15, 17, 18, 25, 30, 32, 39, 61, 63, 64, 66, 72, 76, 89
  Amount of, 92
  Electricity for, 97

INDEX.

Pumps—
  Ancient, 92, 93
  Bascule, 92
  Box, 93
  Bucket-wheel, 92
  Capacity, 77, 78, 93, 94, 99, 101, 135, 148
  Centrifugal, 15, 17, 18, 30, 39, 61, 63, 66, 77, 78, 90, 96, 97, 98, 99, 100, 148, 149
  Chapelet, 92, 93
  Chattanooga plant, 99
  Cost of, 93, 148
  Direct connected, 97
  Double suction, 100
  Dredging, 78, 99, 101, 148
  Edward's cataract, 101
  Efficiency of, 96
  Ejector for priming, 101, 148
  Electric, 97
  Forth bridge, 99
  Furnished, 134
  Gate valve primer, 101
  German high test, 97
  Hand, 93
  Heald & Sisco, 97, 148
  Lift of, 96, 100, 148
  Location, best, 100
  Maslin, 96
  Piston of dredging, 101
  Primers, 100, 148
  Pulsometer, 30, 32, 90, 94, 95
  Reciprocating, 96
  Sampling, 121, 122
  Siphon, 93
  Speed for, 149
  Strainers for, 98
  Suction details, 98
  Suction pipe, 98, 100, 148
  Sump or well for, 60, 99
  Tests of, 96, 97
  Weight of, 148
  Wooden, 93
Quality of—
  Cement, 114, 136, 142, 145
  Timber, 138, 139, 146
Quicksand, 67, 135
Rammer for puddle, 28, 77
Red river, U. S., 111
Rejection of material, 137, 139
Removal of—
  Coffer-dam, 20, 33, 74, 134, 147
  Grillage, 74
  Piers, 72, 74
  Piles, 50, 73
  Test core, 122
Republican river, Kan., 20
Requirements, War Dep't, 120
Rhine river, 5
Rings for piles, 70
Rip-rap, 26, 72, 83
Rivers, navigable, 120
River—
  Aa, Russia, 83
  Arkansas, 20, 70

River—*Continued*.
  Bear, Can., 83
  Charles, Boston, 63, 110
  Clyde, Scotland, 6
  Coosa, Ala., 74, 110, 117
  Danube, 9, 40
  Huron, Mich., 59
  Illinois, 62
  Kanawah, W. Va., 14, 117
  Kankakee, Ill., 62
  Karun, Persia, 1
  Little Bras d'Or, 82
  Missouri, 127
  Mississippi, 20, 32, 67
  Ohio, 15
  Parnitz, Germany, 72
  Potomac, 63
  Red, 111
  Republican, Kan., 20
  Rhine, 5
  Sault Ste. Marie, 28
  Schuylkill, Pa., 74
  Scioto, Ohio, 13
  Soane, France, 109
  St. Lawrence, 18
  Tennessee, 66
  Thames, 4, 77
  Western, U. S., 17, 18, 20
Riveting—
  Boiler, 80, 82, 147
  Water-tight, 80, 82, 147
Roads, construction, 3
Robinson, coffer-dam, 24
Rock—
  Bottom, *see* Bottom
  Crevices, 25, 28, 39, 66, 89, 107
  Joint with sheet piles, 72
  Ledges, 86
  Stepping of, 86, 107
Rock-drill pile-driver, 14
Rod bracing, 30, 38, 67, 134
Roman foundations, 1
Rubble masonry, 86, 90
Rules laying concrete, 113, 138, 143, 144, 145
Sacked puddle, *see* Puddle
Sacks used, 26, 28, 33, 36, 64, 66, 76
Samples of borings, 121, 122
Sampling pumps, 121, 122
Sand (*see* Bottom)—
  For concrete, 111, 115, 136, 143, 145
  Digger, 102
  Drift in, 18
  Pump, 121, 122
  Shells in, 77
Sault Ste. Marie river, 28
Sawing off piles, 50, 83
Schuylkill river, 74
Scioto river, Ohio, 13
Scow, pile pulling, 50
Scow, *see* Pile-driver
Scraper, 13
Sec'y of War, approval, 120
Setting time, concrete, 110, 114, 143, 144, 145

Sewer coffer-dam, 60, 61
Shale, 60
Sheet pile coffer-dam, *see* Coffer-dam
Sheet piles—
    Calculations, 54, 56
    Clamps for, 57
    Close, 135
    Compound, 52, 53, 61, 66
    Driver for, 14, 41
    Early use, 4, 6, 7
    Forms of, 50, 51
    Guides, 57
    Leak remedy, 12, 28
    Metal, 91
    Old, 72
    Plank, 35, 50, 59, 60, 66, 67, 77, 134
    Pointing, 24, 50, 51, 57, 60
    Rock joint with, 72
    Shoes for, 50, 51, 52
    Slanting, 60, 61
    Square, 8, 10, 24, 51, 60, 77
    Thickness, 53, 54, 56
    Tongue and groove, 20, 24, 51, 60, 71, 76, 77, 78
    V shape, 20, 51
    Wakefield, 14, 20, 53, 61, 62, 63, 64, 141
Shells in bottom, 77
Shocks of waves, 83, 89
Shoe for—
    Piles, 50, 53
    Pipe, 122
    Sheet piles, 50, 51, 52
Shoring, *see* Bracing
Silt bottom, 33
Sinking coffer-dam, 20, 76
Siphon, 93
Site of bridge, 120
Site, examination of, 120, 124
Slanting piles, 60, 61
Sledge for pile-driver, 40
Sluice, 7, 87
Soane river, France, 109
Soapstone bottom, 20
Sounding rod, 86
Soundings, 86, 120, 121
Spans—
    Cableway, 117, 118
    Economic formula, 125
    Economic length, 124, 125
    Length of, 124
    Relation to piers, 1, 124
Spacing of piers, 120, 124, 125
Spears for drill rope, 121
Specification for—
    Coffer-dam, *see* Coffer-dam
    Cement, *see* Cement
    Foundation, 141, 146
    Masonry, 136, 137, 138
    Piles, 135, 141, 146
    Steel coffer-dam, 147
    Timber, 138, 139, 146
Speed for pumps, 149
Splicing coffer-dams, 67, 147

Springs in bottom, 32
Spuds, 75
Staging to locate pier, 86
Starlings, 10, 18, 26, 66, 125, 129, 130, 131
Steel coffer-dam, 83, 86, 147
Steel shells, 80, 82, 83
St. Lawrence river, 18
Stock rammer, 12, 28, 77
Stone—
    Concretes., 111, 115, 143, 145
    Defects in, 137
    Dressing, 137
    Protection, 83, 135
    Quality of, 136
    Rubble, 137
    Samples, 137
    Substitute for, 80
Strainer for pump, 98
Struts, *see* Timber
Suction-pipe, 98, 148
Sump for pump 60, 99
Suspension bridge tower, 8
Swab, 121
Swift water, *see* Water
Tables of—
    Pumps, centrifugal, 148
    Pumps, dredging, 148
    Pump speed, 149
    Hoisting engines, 149, 150
Tarpaulin used, 30, 32, 33, 35, 36, 65
Tennessee river, 66
Test borings, *see* Borings
Tests of—
    Cement, 115, 136, 142, 145
    Pumps, 96, 97
Thames river, 4, 77
Thickness of—
    Cylinders, 84
    Sheet piles, 53, 54, 56
Tide coffer-dam, 8, 53, 63, 86, 89
Tide, *see* Water
Timber—
    Casing, 5
    Framing, 135
    Piles, *see* Piles
        see sheet piles
    Price for, 139, 146
    Scarcity of, 79
    Specifications, 138, 139, 146
    Struts, 7, 30, 33, 38, 39, 56, 64
    Water-soaked, 33
Tongs—
    Pebble, 124
    Pile-driver, 43
Tongue and groove, *see* Sheet piles
Tools furnished, 134
Topeka coffer dam, 78
Trezzo arch, 1
Tripod for test boring, 121
Tube for—
    Concreting, 108, 110
    Drilling, 121
    Pier, 33, 80, 82, 83, 87

Van Duzen jet, 93
Velocity of water, 129, 130, 131
Victoria, B. C., docks, 77
Voids in masonry, 136
V shape pile joint, 20, 51
Wakefield sheet piles, *see* Sheet piles
Wales—
  Calculation of, 56
  Ordinary, 7, 35, 59, 134
War Dep't requirements, 120
Washers for leaks, 30
Water—
  For concrete, 113, 114, 143
  Deep, 8, 10, 63
  Eddies, 129, 130, 131
  High, 118, 119, 120
  Jet, 70
  Low, 120
  Pressure, 20, 25, 28, 32, 35, 54, 85, 89

Water—*Continued.*
  Shallow, 1, 5, 13
  Shock of waves, 83, 89
  Soaked timber, 33
  Swift, 10, 26, 32, 107
  Tide, 8, 86
  Velocity, 129, 130, 131
Water-tight compartments, 76
Water-tight riveting, 80, 82, 147
Waterway, 120, 129
Weight of—
  Hoist engines, 149, 150
  Pumps, 148
Well driller, 121
Well for pump, 60, 99
Wells in concrete, 114
Wellington pile formula, 49
Western, U. S., rivers, 17, 18, 20
Windlass pile-driver, 40
Wooden pump, 93

Wakefield Triple=lap Sheet Piling.

This sheeting matches perfectly; can be made as wanted at the work, of any available sound lumber.; stands driving without an equal, and <u>stops water absolutely</u> at such trifling cost that it invariably proves satisfactory.

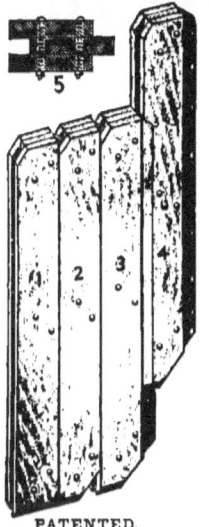

PATENTED.

Royalty charge, 25 cents per foot of completed work; i.e., a coffer-dam 25' × 50', sheeted on four sides, amounts to 150' @ 25c. = $37.50.

USED BY ARMY ENGINEERS, RAILWAY ENGINEERS, CITY ENGINEERS, CONTRACTORS, ETC.

WAKEFIELD SHEET=PILING CO., 770 The Rookery, Chicago, Ill.

Heald & Sisco Centrifugal Pumps

FOR

COFFER-DAM WORK AND GENERAL CONTRACTING.

Our New Special Hydraulic Dredging and Sand Pump embodies the latest improvements suggested by the experience of contractors all over the country. We confidently recommend it, therefore, as the best and most efficient pump manufactured. Some of the new features are: an improved piston; a disk on the suction side, which can be easily removed to get at the inside without dismantling the whole pump; a shaft-bearing in the pump-shell adjustable to take up wear. We also furnish, if desired, steel wearing plates and cast-steel liners for the inside of the shell, which lengthen the life of the pump considerably. Write for catalogue containing full particulars, and ask for testimonials and prices.

MORRIS MACHINE WORKS,

New York Office, 39-41 Cortlandt St.   *BALDWINSVILLE, N. Y.*

Chicago Agents, HENION & HUBBELL, 61-69 North Jefferson St.

ATLAS PORTLAND CEMENT

IS THE

Standard American Brand.

Used by the Leading Engineers throughout the United States.

SPECIALLY ADAPTED FOR

HEAVY MASONRY WORK.

SEND FOR PAMPHLETS.

ATLAS CEMENT CO.,
143 LIBERTY STREET,   NEW YORK CITY.

LIDGERWOOD Hoisting Engines

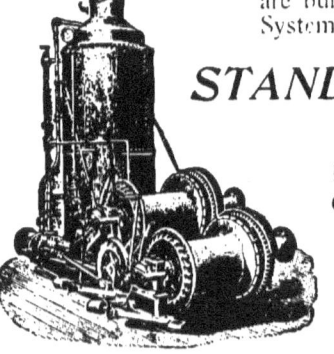

are built to gauge on the Duplicate Part System. Quick Delivery Assured.

STANDARD FOR QUALITY AND DUTY.

FOR PILE DRIVING,
Dock and Bridge Building,
Contractors, and General Hoisting.

OVER **13,000** IN USE.

CABLEWAYS,
HOISTING AND CONVEYING DEVICES, STEAM AND ELECTRIC HOISTS, for General Contract Work.

SEND FOR LATEST CATALOGUE.

LIDGERWOOD MFG. CO., 96 Liberty St., New York.

SHORT-TITLE CATALOGUE

OF THE

PUBLICATIONS

OF

JOHN WILEY & SONS,

NEW YORK.

LONDON: CHAPMAN & HALL, LIMITED.

ARRANGED UNDER SUBJECTS.

Descriptive circulars sent on application.
Books marked with an asterisk are sold at *net* prices only.
All books are bound in cloth unless otherwise stated.

AGRICULTURE.

CATTLE FEEDING—DAIRY PRACTICE—DISEASES OF ANIMALS—GARDENING, ETC.

Armsby's Manual of Cattle Feeding....................12mo,	$1	75
Downing's Fruit and Fruit Trees........................8vo,	5	00
Grotenfelt's The Principles of Modern Dairy Practice. (Woll.) 12mo,	2	00
Kemp's Landscape Gardening..........................12mo,	2	50
Lloyd's Science of Agriculture..........................8vo,	4	00
Loudon's Gardening for Ladies. (Downing.)............12mo,	1	50
Steel's Treatise on the Diseases of the Dog................8vo,	3	50
" Treatise on the Diseases of the Ox..................8vo,	6	00
Stockbridge's Rocks and Soils........ ....8vo,	2	50
Woll's Handbook for Farmers and Dairymen............12mo,	1	50

ARCHITECTURE.

BUILDING—CARPENTRY—STAIRS—VENTILATION, ETC.

Berg's Buildings and Structures of American Railroads.....4to,	7	50
Birkmire's American Theatres—Planning and Construction.8vo,	3	00
" Architectural Iron and Steel....................8vo,	3	50
" Compound Riveted Girders....................8vo,	2	00
" Skeleton Construction in Buildings 8vo,	3	00
" Planning and Construction of High Office Buildings. 8vo,	3	50

Carpenter's Heating and Ventilating of Buildings	8vo,	$3 00
Downing, Cottages	8vo,	2 50
" Hints to Architects	8vo,	2 00
Freitag's Architectural Engineering	8vo,	2 50
Gerhard's Sanitary House Inspection	16mo,	1 00
" Theatre Fires and Panics	12mo,	1 50
Hatfield's American House Carpenter	8vo,	5 00
Holly's Carpenter and Joiner	18mo,	75
Kidder's Architect and Builder's Pocket-book	Morocco flap,	4 00
Merrill's Stones for Building and Decoration	8vo,	5 00
Monckton's Stair Building—Wood, Iron, and Stone	4to,	4 00
Wait's Engineering and Architectural Jurisprudence	8vo,	6 00
	Sheep,	6 50
Worcester's Small Hospitals—Establishment and Maintenance, including Atkinson's Suggestions for Hospital Architecture	12mo,	1 25
World's Columbian Exposition of 1893	4to,	2 50

ARMY, NAVY, Etc.

MILITARY ENGINEERING—ORDNANCE—PORT CHARGES—LAW, ETC.

Bourne's Screw Propellers	4to,	5 00
Bruff's Ordnance and Gunnery	8vo,	6 00
Bucknill's Submarine Mines and Torpedoes	8vo,	4 00
Chase's Screw Propellers	8vo,	3 00
Cooke's Naval Ordnance	8vo,	12 50
Cronkhite's Gunnery for Non-com. Officers	18mo, morocco,	2 00
Davis's Treatise on Military Law	8vo,	7 00
	Sheep,	7 50
De Brack's Cavalry Outpost Duties. (Carr.)	18mo, morocco,	2 00
Dietz's Soldier's First Aid	12mo, morocco,	1 25
*Dredge's Modern French Artillery	4to, half morocco,	20 00
" Record of the Transportation Exhibits Building, World's Columbian Exposition of 1893	4to, half morocco,	10 00
Durand's Resistance and Propulsion of Ships	8vo,	5 00
Dyer's Light Artillery	12mo,	3 00
Hoff's Naval Tactics	8vo,	1 50
Hunter's Port Charges	8vo, half morocco,	13 00
Ingalls's Ballistic Tables	8vo,	1 50

Ingalls's Handbook of Problems in Direct Fire............8vo,	4 00
Mahan's Advanced Guard..............................18mo,	$1 50
Mahan's Permanent Fortifications. (Mercur.).8vo, half morocco,	7 50
Mercur's Attack of Fortified Places....................12mo,	2 00
Mercur's Elements of the Art of War....................8vo,	4 00
Metcalfe's Ordnance and Gunnery...........12mo, with Atlas,	5 00
Murray's A Manual for Courts-Martial........18mo, morocco,	1 50
" Infantry Drill Regulations adapted to the Springfield Rifle, Caliber .45....................18mo, paper,	15
Phelps's Practical Marine Surveying......................8vo,	2 50
Powell's Army Officer's Examiner......................12mo,	4 00
Reed's Signal Service..	50
Sharpe's Subsisting Armies..................18mo, morocco,	1 50
Very's Navies of the World...............8vo, half morocco,	3 50
Wheeler's Siege Operations..............................8vo,	2 00
Winthrop's Abridgment of Military Law................12mo,	2 50
Woodhull's Notes on Military Hygiene.........12mo, morocco,	2 50
Young's Simple Elements of Navigation..12mo, morocco flaps,	2 50
" " " " " first edition........	1 00

ASSAYING.
SMELTING—ORE DRESSING—ALLOYS, ETC.

Fletcher's Quant. Assaying with the Blowpipe..12mo, morocco,	1 50
Furman's Practical Assaying.............................8vo,	3 00
Kunhardt's Ore Dressing................................8vo,	1 50
* Mitchell's Practical Assaying. (Crookes.)..............8vo,	10 00
O'Driscoll's Treatment of Gold Ores....................8vo,	2 00
Ricketts and Miller's Notes on Assaying.................8vo,	3 00
Thurston's Alloys, Brasses, and Bronzes...............8vo,	2 50
Wilson's Cyanide Processes...........................12mo,	1 50
" The Chlorination Process....................12mo,	1 50

ASTRONOMY.
PRACTICAL, THEORETICAL, AND DESCRIPTIVE.

Craig's Azimuth..4to,	3 50
Doolittle's Practical Astronomy..........................8vo,	4 00
Gore's Elements of Geodesy..............................8vo,	2 50
Michie and Harlow's Practical Astronomy................8vo,	3 00
White's Theoretical and Descriptive Astronomy..........12mo,	2 00

BOTANY.

Gardening for Ladies, Etc.

Baldwin's Orchids of New England.................8vo,	$1	50
Loudon's Gardening for Ladies. (Downing.)...........12mo,	1	50
Thomé's Structural Botany.................................18mo,	2	25
Westermaier's General Botany. (Schneider.) 8vo,	2	00

BRIDGES, ROOFS, Etc.

Cantilever—Draw—Highway—Suspension.

(See also Engineering, p. 6.)

Boller's Highway Bridges.................................8vo,	2	00
* " The Thames River Bridge.................4to, paper,	5	00
Burr's Stresses in Bridges................................8vo,	3	50
Crehore's Mechanics of the Girder........................8vo,	5	00
Dredge's Thames Bridges....................7 parts, per part,	1	25
Du Bois's Stresses in Framed Structures..................4to,	10	00
Foster's Wooden Trestle Bridges.........................4to,	5	00
Greene's Arches in Wood, etc............................8vo,	2	50
" Bridge Trusses.................................8vo,	2	50
" Roof Trusses...................................8vo,	1	25
Howe's Treatise on Arches 8vo,	4	00
Johnson's Modern Framed Structures......................4to,	10	00
Merriman & Jacoby's Text-book of Roofs and Bridges. Part I., Stresses......................................8vo,	2	50
Merriman & Jacoby's Text-book of Roofs and Bridges. Part II., Graphic Statics..............................8vo,	2	50
Merriman & Jacoby's Text-book of Roofs and Bridges. Part III., Bridge Design..............................8vo,	2	50
Merriman & Jacoby's Text-book of Roofs and Bridges. Part IV., Continuous, Draw, Cantilever, Suspension, and Arched Bridges.....................................8vo,	2	50
* Morison's The Memphis Bridge.................Oblong 4to,	10	00
Waddell's Iron Highway Bridges........ ....8vo,	4	00
" De Pontibus (a Pocket-book for Bridge Engineers).		
Wood's Construction of Bridges and Roofs...............8vo,	2	00
Wright's Designing of Draw Spans...................... 8vo,	2	50

CHEMISTRY.

QUALITATIVE—QUANTITATIVE—ORGANIC—INORGANIC, ETC.

Adriance's Laboratory Calculations.....................12mo,	$1 25
Allen's Tables for Iron Analysis..........................8vo,	3 00
Austen's Notes for Chemical Students....................12mo,	1 50
Bolton's Student's Guide in Quantitative Analysis.........8vo,	1 50
Classen's Analysis by Electrolysis. (Herrick and Boltwood.).8vo,	3 00
Crafts's Qualitative Analysis. (Schaeffer.)..............12mo,	1 50
Drechsel's Chemical Reactions. (Merrill.)...............12mo,	1 25
Fresenius's Quantitative Chemical Analysis. (Allen.).......8vo,	6 00
" Qualitative " " (Johnson.).....8vo,	3 00
" " " " (Wells) Trans. 16th.	
German Edition..................................8vo,	5 00
Fuerte's Water and Public Health.......................12mo,	1 50
Gill's Gas and Fuel Analysis............................12mo,	1 25
Hammarsten's Physiological Chemistry. (Mandel.).........8vo,	4 00
Helm's Principles of Mathematical Chemistry. (Morgan).12mo,	1 50
Kolbe's Inorganic Chemistry............................12mo,	1 50
Ladd's Quantitative Chemical Analysis..................12mo.	
Landauer's Spectrum Analysis. (Tingle.)..................8vo,	3 00
Mandel's Bio-chemical Laboratory.......................12mo,	1 50
Mason's Water-supply...................................8vo,	5 00
" Analysis of Potable Water. (*In the press*.)	
Miller's Chemical Physics................................8vo,	2 00
Mixter's Elementary Text-book of Chemistry............12mo,	1 50
Morgan's The Theory of Solutions and its Results.......12mo,	1 00
Nichols's Water-supply (Chemical and Sanitary)...........8vo,	2 50
O'Brine's Laboratory Guide to Chemical Analysis..........8vo,	2 00
Perkins's Qualitative Analysis..........................12mo,	1 00
Pinner's Organic Chemistry. (Austen.)..................12mo,	1 50
Poole's Calorific Power of Fuels..........................8vo,	3 00
Ricketts and Russell's Notes on Inorganic Chemistry (Non-metallic)................. Oblong 8vo, morocco,	75
Ruddiman's Incompatibilities in Prescriptions............8vo,	2 00
Schimpf's Volumetric Analysis..........................12mo,	2 50
Spencer's Sugar Manufacturer's Handbook.12mo, morocco flaps,	2 00
" Handbook for Chemists of Beet Sugar House. 12mo, morocco,	3 00

Stockbridge's Rocks and Soils................8vo,	$2	50
Troilius's Chemistry of Iron................8vo,	2	00
Wells's Qualitative Analysis................12mo.		
Wiechmann's Chemical Lecture Notes........12mo,	3	00
" Sugar Analysis................8vo,	2	50
Wulling's Inorganic Phar. and Med. Chemistry......12mo,	2	00

DRAWING.
ELEMENTARY—GEOMETRICAL—TOPOGRAPHICAL.

Hill's Shades and Shadows and Perspective..........8vo,	2	00
MacCord's Descriptive Geometry................8vo,	3	00
MacCord's Kinematics................8vo,	5	00
" Mechanical Drawing................8vo,	4	00
Mahan's Industrial Drawing. (Thompson.)......2 vols., 8vo,	3	50
Reed's Topographical Drawing. (H. A.)................4to,	5	00
Reid's A Course in Mechanical Drawing................8vo.	2	00
" Mechanical Drawing and Elementary Mechanical Design. 8vo.		
Smith's Topographical Drawing. (Macmillan.)........8vo,	2	50
Warren's Descriptive Geometry............2 vols., 8vo,	3	50
" Drafting Instruments................12mo,	1	25
" Free-hand Drawing................12mo,	1	00
" Higher Linear Perspective................8vo,	3	50
" Linear Perspective................12mo,	1	00
" Machine Construction........2 vols., 8vo,	7	50
" Plane Problems................12mo,	1	25
" Primary Geometry................12mo,		75
" Problems and Theorems................8vo,	2	50
" Projection Drawing................12mo,	1	50
" Shades and Shadows................8vo,	3	00
" Stereotomy—Stone Cutting................8vo,	2	50
Whelpley's Letter Engraving................12mo,	2	00

ELECTRICITY AND MAGNETISM.
ILLUMINATION—BATTERIES—PHYSICS.

Anthony and Brackett's Text-book of Physics (Magie)....8vo,	4	00
Barker's Deep-sea Soundings................8vo,	2	00
Benjamin's Voltaic Cell................8vo,	3	00
" History of Electricity................8vo	3	00

Cosmic Law of Thermal Repulsion 18mo,	$	75
Crehore and Squier's Experiments with a New Polarizing Photo-Chronograph. 8vo,	3	00
* Dredge's Electric Illuminations;...2 vols., 4to, half morocco,	25	00
" " " Vol. II..................4to,	7	50
Gilbert's De magnete. (Mottelay.)......................8vo,	2	50
Holman's Precision of Measurements....................8vo,	2	00
Michie's Wave Motion Relating to Sound and Light,.......8vo,	4	00
Morgan's The Theory of Solutions and its Results........12mo,	1	00
Niaudet's Electric Batteries. (Fishback.)...............12mo,	2	50
Reagan's Steam and Electrical Locomotives............12mo,	2	00
Thurston's Stationary Steam Engines for Electric Lighting Purposes......................................12mo,	1	50
Tillman's Heat..8vo,	1	50

ENGINEERING.

Civil—Mechanical—Sanitary, Etc.

(*See also* Bridges, p. 4; Hydraulics, p. 8; Materials of Engineering, p. 9; Mechanics and Machinery, p. 11; Steam Engines and Boilers, p. 14.)

Baker's Masonry Construction........................8vo,	5	00
Baker's Surveying Instruments........................12mo,	3	00
Black's U. S. Public Works............................4to,	5	00
Brook's Street Railway Location..............12mo, morocco,	1	50
Butts's Engineer's Field-book.................12mo, morocco,	2	50
Byrne's Highway Construction.........................8vo,	7	50
Carpenter's Experimental Engineering 8vo,	6	00
Church's Mechanics of Engineering—Solids and Fluids....8vo,	6	00
" Notes and Examples in Mechanics..............8vo,	2	00
Crandall's Earthwork Tables 8vo,	1	50
Crandall's The Transition Curve..............12mo, morocco,	1	50
* Dredge's Penn. Railroad Construction, etc... Folio, half mor.,	20	00
* Drinker's Tunnelling....................4to, half morocco,	25	00
Eissler's Explosives—Nitroglycerine and Dynamite........8vo,	4	00
Gerhard's Sanitary House Inspection....................16mo,	1	00
Godwin's Railroad Engineer's Field-book,12mo, pocket-bk. form,	2	50
Gore's Elements of Geodesy..........................8vo,	2	50
Howard's Transition Curve Field-book.....12mo, morocco flap,	1	50
Howe's Retaining Walls (New Edition.)................12mo,	1	25

Hudson's Excavation Tables. Vol. II..................8vo,	$1 00
Hutton's Mechanical Engineering of Power Plants........8vo,	5 00
Johnson's Materials of Construction.....................8vo,	6 00
Johnson's Stadia Reduction Diagram..Sheet, 22½ × 28½ inches,	50
" Theory and Practice of Surveying..............8vo,	4 00
Kent's Mechanical Engineer's Pocket-book.....12mo, morocco,	5 00
Kiersted's Sewage Disposal........................12mo,	1 25
Kirkwood's Lead Pipe for Service Pipe..................8vo,	1 50
Mahan's Civil Engineering. (Wood.)....................8vo,	5 00
Merriman and Brook's Handbook for Surveyors....12mo, mor.,	2 00
Merriman's Geodetic Surveying..........................8vo,	2 00
" Retaining Walls and Masonry Dams..........8vo,	2 00
Mosely's Mechanical Engineering. (Mahan.)..............8vo,	5 00
Nagle's Manual for Railroad Engineers........12mo, morocco,	3 00
Patton's Civil Engineering.............................8vo,	7 50
" Foundations...................................8vo,	5 00
Rockwell's Roads and Pavements in France.........12mo,	1 25
Ruffner's Non-tidal Rivers 8vo,	1 25
Searles's Field Engineering..............12mo, morocco flaps,	3 00
" Railroad Spiral ....12mo, morocco flaps,	1 50
Siebert and Biggin's Modern Stone Cutting and Masonry...8vo,	1 50
Smith's Cable Tramways.................................4to,	2 50
" Wire Manufacture and Uses.....................4to,	3 00
Spalding's Roads and Pavements.......................12mo,	2 00
" Hydraulic Cement...........................12mo,	2 00
Thurston's Materials of Construction 8vo,	5 00
* Trautwine's Civil Engineer's Pocket-book...12mo, mor. flaps,	5 00
* " Cross-section.........................Sheet,	25
* " Excavations and Embankments............8vo,	2 00
* " Laying Out Curves...........12mo, morocco,	2 50
Waddell's De Pontibus (A Pocket-book for Bridge Engineers). 12mo, morocco,	3 00
Wait's Engineering and Architectural Jurisprudence.......8vo,	6 00
Sheep,	6 50
Warren's Stereotomy—Stone Cutting.....................8vo,	2 50
Webb's Engineering Instruments.............12mo, morocco,	1 00
Wegmann's Construction of Masonry Dams..............4to,	5 00
Wellington's Location of Railways.......................8vo,	5 00

Wheeler's Civil Engineering................................8vo,	$4	00
Wolff's Windmill as a Prime Mover......................8vo,	3	00

HYDRAULICS.

WATER-WHEELS—WINDMILLS—SERVICE PIPE—DRAINAGE, ETC.
(*See also* ENGINEERING, p. 6.)

Bazin's Experiments upon the Contraction of the Liquid Vein (Trautwine)..8vo,	2	00
Bovey's Treatise on Hydraulics..............................8vo,	4	00
Coffin's Graphical Solution of Hydraulic Problems.......12mo,	2	50
Ferrel's Treatise on the Winds, Cyclones, and Tornadoes...8vo,	4	00
Fuerte's Water and Public Health.........................12mo,	1	50
Ganguillet & Kutter's Flow of Water. (Hering & Trautwine.).8vo,	4	00
Hazen's Filtration of Public Water Supply................8vo,	2	00
Herschel's 115 Experiments 8vo,	2	00
Kiersted's Sewage Disposal...............................12mo,	1	25
Kirkwood's Lead Pipe for Service Pipe 8vo,	1	50
Mason's Water Supply.....................................8vo,	5	00
Merriman's Treatise on Hydraulics........................8vo,	4	00
Nichols's Water Supply (Chemical and Sanitary)...........8vo,	2	50
Ruffner's Improvement for Non-tidal Rivers...............8vo,	1	25
Wegmann's Water Supply of the City of New York.......4to,	10	00
Weisbach's Hydraulics. (Du Bois.)........................8vo,	5	00
Wilson's Irrigation Engineering...........................8vo,	4	00
Wolff's Windmill as a Prime Mover........................8vo,	3	00
Wood's Theory of Turbines................................8vo,	2	50

MANUFACTURES.

ANILINE—BOILERS—EXPLOSIVES—IRON—SUGAR—WATCHES—
WOOLLENS, ETC.

Allen's Tables for Iron Analysis...........................8vo,	3	00
Beaumont's Woollen and Worsted Manufacture.........12mo,	1	50
Bolland's Encyclopædia of Founding Terms.............12mo,	3	00
" The Iron Founder..................................12mo,	2	50
" " " " Supplement......................12mo,	2	50
Booth's Clock and Watch Maker's Manual................12mo,	2	00
Bouvier's Handbook on Oil Painting.....................12mo,	2	00
Eissler's Explosives, Nitroglycerine and Dynamite........8vo,	4	00
Ford's Boiler Making for Boiler Makers..................18mo,	1	00
Metcalfe's Cost of Manufactures...........................8vo,	5	00

Metcalf's Steel—A Manual for Steel Users............12mo,	$2	00
Reimann's Aniline Colors. (Crookes.)..................8vo,	2	50
* Reisig's Guide to Piece Dyeing.......................8vo,	25	00
Spencer's Sugar Manufacturer's Handbook....12mo, mor. flap,	2	00
" Handbook for Chemists of Beet Houses. 12mo, mor. flap,	3	00
Svedelius's Handbook for Charcoal Burners............12mo,	1	50
The Lathe and Its Uses................................8vo,	6	00
Thurston's Manual of Steam Boilers....................8vo,	5	00
Walke's Lectures on Explosives........................8vo,	4	00
West's American Foundry Practice.....................12mo,	2	50
" Moulder's Text-book 12mo,	2	50
Wiechmann's Sugar Analysis...........................8vo,	2	50
Woodbury's Fire Protection of Mills..................8vo,	2	50

MATERIALS OF ENGINEERING.

STRENGTH—ELASTICITY—RESISTANCE, ETC.

(*See also* ENGINEERING, p. 6.)

Baker's Masonry Construction..........................8vo,	5	00
Beardslee and Kent's Strength of Wrought Iron.........8vo,	1	50
Bovey's Strength of Materials.........................8vo,	7	50
Burr's Elasticity and Resistance of Materials.........8vo,	5	00
Byrne's Highway Construction.........................8vo,	5	00
Carpenter's Testing Machines and Methods of Testing Materials.		
Church's Mechanics of Engineering—Solids and Fluids.....8vo,	6	00
Du Bois's Stresses in Framed Structures..............4to,	10	00
Hatfield's Transverse Strains.........................8vo,	5	00
Johnson's Materials of Construction...................8vo,	6	00
Lanza's Applied Mechanics............................8vo,	7	50
" Strength of Wooden Columns ....8vo, paper,		50
Merrill's Stones for Building and Decoration..........8vo,	5	00
Merriman's Mechanics of Materials.....................8vo,	4	00
" Strength of Materials................... 12mo,	1	00
Patton's Treatise on Foundations......................8vo,	5	00
Rockwell's Roads and Pavements in France.............12mo,	1	25
Spalding's Roads and Pavements.......................12mo,	2	00
Thurston's Materials of Construction..................8vo,	5	00

Thurston's Materials of Engineering..............3 vols., 8vo,	$8 00
Vol. I., Non-metallic 8vo,	2 00
Vol. II., Iron and Steel............................8vo,	3 50
Vol. III., Alloys, Brasses, and Bronzes..............8vo,	2 50
Weyrauch's Strength of Iron and Steel. (Du Bois.).........8vo,	1 50
Wood's Resistance of Materials............................8vo,	2 00

MATHEMATICS.
CALCULUS—GEOMETRY—TRIGONOMETRY, ETC.

Baker's Elliptic Functions...............................8vo,	1 50
Ballard's Pyramid Problem.............. ....8vo,	1 50
Barnard's Pyramid Problem...........................8vo,	1 50
Bass's Differential Calculus...........................12mo,	4 00
Brigg's Plane Analytical Geometry.....................12mo,	1 00
Chapman's Theory of Equations........................12mo,	1 50
Chessin's Elements of the Theory of Functions.	
Compton's Logarithmic Computations...................12mo,	1 50
Craig's Linear Differential Equations................. ...8vo,	5 00
Davis's Introduction to the Logic of Algebra.............8vo,	1 50
Halsted's Elements of Geometry........... ....8vo,	1 75
" Synthetic Geometry............................8vo,	1 50
Johnson's Curve Tracing...............................12mo,	1 00
" Differential Equations—Ordinary and Partial.....8vo,	3 50
" Integral Calculus............................12mo,	1 50
" " " Unabridged.	
" Least Squares...............................12mo,	1 50
Ludlow's Logarithmic and Other Tables. (Bass.).........8vo,	2 00
" Trigonometry with Tables. (Bass.).............8vo,	3 00
Mahan's Descriptive Geometry (Stone Cutting).... ....8vo,	1 50
Merriman and Woodward's Higher Mathematics......... ..8vo,	5 00
Merriman's Method of Least Squares 8vo,	2 00
Parker's Quadrature of the Circle 8vo,	2 50
Rice and Johnson's Differential and Integral Calculus,	
2 vols. in 1, 12mo,	2 50
" Differential Calculus....................8vo,	3 00
" Abridgment of Differential Calculus....8vo,	1 50
Searles's Elements of Geometry. 8vo,	1 50
Totten's Metrology.....................................8vo,	2 50
Warren's Descriptive Geometry...................2 vols., 8vo,	3 50
" Drafting Instruments........................12mo,	1 25
" Free-hand Drawing.........................12mo,	1 00
" Higher Linear Perspective....................8vo,	3 50
" Linear Perspective..........................12mo,	1 00
" Primary Geometry..........................12mo,	75

Warren's Plane Problems 12mo,	$1	25
" Plane Problems 12mo,	1	25
" Problems and Theorems 8vo,	2	50
" Projection Drawing 12mo,	1	50
Wood's Co-ordinate Geometry 8vo,	2	00
" Trigonometry 12mo,	1	00
Woolf's Descriptive Geometry ....Royal 8vo,	3	00

MECHANICS—MACHINERY.

TEXT-BOOKS AND PRACTICAL WORKS.

(*See also* ENGINEERING, p. 6.)

Baldwin's Steam Heating for Buildings 12mo,	2	50
Benjamin's Wrinkles and Recipes 12mo,	2	00
Carpenter's Testing Machines and Methods of Testing Materials ....8vo.		
Chordal's Letters to Mechanics 12mo,	2	00
Church's Mechanics of Engineering 8vo,	6	00
" Notes and Examples in Mechanics 8vo,	2	00
Crehore's Mechanics of the Girder 8vo,	5	00
Cromwell's Belts and Pulleys 12mo,	1	50
" Toothed Gearing 12mo,	1	50
Compton's First Lessons in Metal Working 12mo,	1	50
Dana's Elementary Mechanics 12mo,	1	50
Dingey's Machinery Pattern Making 12mo,	2	00
Dredge's Trans. Exhibits Building, World Exposition, 4to, half morocco,	10	00
Du Bois's Mechanics. Vol. I., Kinematics 8vo,	3	50
" " Vol. II., Statics 8vo,	4	00
" " Vol. III., Kinetics 8vo,	3	50
Fitzgerald's Boston Machinist 18mo,	1	00
Flather's Dynamometers 12mo,	2	00
" Rope Driving 12mo,	2	00
Hall's Car Lubrication 12mo,	1	00
Holly's Saw Filing 18mo,		75
Johnson's Theoretical Mechanics. An Elementary Treatise. (*In the press.*)		
Jones Machine Design. Part I., Kinematics 8vo,	1	50
" " " Part II., Strength and Proportion of Machine Parts.		
Lanza's Applied Mechanics 8vo,	7	50
MacCord's Kinematics 8vo,	5	00
Merriman's Mechanics of Materials 8vo,	4	00
Metcalfe's Cost of Manufactures 8vo,	5	00
Michie's Analytical Mechanics 8vo,	4	00

Mosely's Mechanical Engineering. (Mahan.)............8vo,	$5	00
Richards's Compressed Air............................12mo,	1	50
Robinson's Principles of Mechanism................ 8vo,	3	00
Smith's Press-working of Metals......................8vo,	3	00
The Lathe and Its Uses............................... 8vo,	6	00
Thurston's Friction and Lost Work....................8vo,	3	00
" The Animal as a Machine..................12mo,	1	00
Warren's Machine Construction................2 vols., 8vo,	7	50
Weisbach's Hydraulics and Hydraulic Motors. (Du Bois.)..8vo,	5	00
" Mechanics of Engineering. Vol. III., Part I., Sec. I. (Klein.)...............................8vo,	5	00
Weisbach's Mechanics of Engineering. Vol. III., Part I., Sec. II. (Klein.)...................................8vo,	5	00
Weisbach's Steam Engines. (Du Bois.)................8vo,	5	00
Wood's Analytical Mechanics..........................8vo,	3	00
" Elementary Mechanics.......................12mo,	1	25
" " " Supplement and Key...........	1	25

METALLURGY.

IRON—GOLD—SILVER—ALLOYS, ETC.

Allen's Tables for Iron Analysis.........................8vo,	3	00
Egleston's Gold and Mercury...........................8vo,	7	50
" Metallurgy of Silver....................... 8vo,	7	50
* Kerl's Metallurgy—Copper and Iron....................8vo,	15	00
* " " Steel, Fuel, etc....................8vo,	15	00
Kunhardt's Ore Dressing in Europe......................8vo,	1	50
Metcalf Steel—A Manual for Steel Users... ....12mo,	2	00
O'Driscoll's Treatment of Gold Ores.....................8vo,	2	00
Thurston's Iron and Steel...............................8vo,	3	50
" Alloys.....................................8vo,	2	50
Wilson's Cyanide Processes...........................12mo,	1	50

MINERALOGY AND MINING.

MINE ACCIDENTS—VENTILATION—ORE DRESSING, ETC.

Barringer's Minerals of Commercial Value....oblong morocco,	2	50
Beard's Ventilation of Mines............... ....12mo,	2	50
Boyd's Resources of South Western Virginia.............8vo,	3	00
" Map of South Western Virginia.....Pocket-book form,	2	00
Brush and Penfield's Determinative Mineralogy.. ....8vo,	3	50
Chester's Catalogue of Minerals........................8vo,	1	25
" " " " ....paper,		50
" Dictionary of the Names of Minerals.............8vo,	3	00
Dana's American Localities of Minerals...................8vo,	1	00

Dana's Descriptive Mineralogy. (E. S.)....8vo, half morocco,	$12	50
" Mineralogy and Petrography (J.D.)............12mo,	2	00
" Minerals and How to Study Them. (E. S.).......12mo,	1	50
" Text-book of Mineralogy. (E. S.)................8vo,	3	50
*Drinker's Tunnelling, Explosives, Compounds, and Rock Drills. 4to, half morocco,	25	00
Egleston's Catalogue of Minerals and Synonyms...........8vo,	2	50
Eissler's Explosives—Nitroglycerine and Dynamite........8vo,	4	00
Goodyear's Coal Mines of the Western Coast............12mo,	2	50
Hussak's Rock-forming Minerals. (Smith.)..............8vo,	2	00
Ihlseng's Manual of Mining.. ....8vo,	4	00
Kunhardt's Ore Dressing in Europe.....................8vo,	1	50
O'Driscoll's Treatment of Gold Ores.....................8vo,	2	00
Rosenbusch's Microscopical Physiography of Minerals and Rocks. (Iddings.)................................8vo,	5	00
Sawyer's Accidents in Mines............... ....8vo,	7	00
Stockbridge's Rocks and Soils...........................8vo,	2	50
Walke's Lectures on Explosives..........................8vo,	4	00
Williams's Lithology....................................8vo,	3	00
Wilson's Mine Ventilation............... ....16mo,	1	25
" Placer Mining....................... ....12mo.		

STEAM AND ELECTRICAL ENGINES, BOILERS, Etc.

STATIONARY—MARINE—LOCOMOTIVE—GAS ENGINES, ETC.

(*See also* ENGINEERING, p. 6.)

Baldwin's Steam Heating for Buildings................12mo,	2	50
Clerk's Gas Engine........................ ....12mo,	4	00
Ford's Boiler Making for Boiler Makers................18mo,	1	00
Hemenway's Indicator Practice.........................12mo,	2	00
Hoadley's Warm-blast Furnace...........................8vo,	1	50
Kneass's Practice and Theory of the Injector............8vo,	1	50
MacCord's Slide Valve.................................8vo,	2	00
* Maw's Marine Engines.................Folio, half morocco,	18	00
Meyer's Modern Locomotive Construction................4to,	10	00
Peabody and Miller's Steam Boilers....... ....8vo,	4	00
Peabody's Tables of Saturated Steam...:................8vo,	1	00
" Thermodynamics of the Steam Engine......... 8vo,	5	00
" Valve Gears for the Steam Engine..............8vo,	2	50
Pray's Twenty Years with the Indicator............Royal 8vo,	2	50
Pupin and Osterberg's Thermodynamics................12mo,	1	25
Reagan's Steam and Electrical Locomotives........ ......12mo,	2	00
Röntgen's Thermodynamics. (Du Bois.)................8vo,	5	00
Sinclair's Locomotive Running.........................12mo,	2	00
Thurston's Boiler Explosion.... ....12mo,	1	50

Thurston's Engine and Boiler Trials.....................8vo,	$5	00
" Manual of the Steam Engine. Part I., Structure and Theory................................8vo,	7	50
" Manual of the Steam Engine. Part II., Design, Construction, and Operation..............8vo,	7	50
2 parts,	12	00
" Philosophy of the Steam Engine.............12mo,		75
" Reflection on the Motive Power of Heat. (Carnot.) • 12mo,	1	50
" Stationary Steam Engines..................12mo,	1	50
" Steam-boiler Construction and Operation.......8vo,	5	00
Spangler's Valve Gears...................................8vo,	2	50
Trowbridge's Stationary Steam Engines...........4to, boards,	2	50
Weisbach's Steam Engine. (Du Bois.)...................8vo,	5	00
Whitham's Constructive Steam Engineering...............8vo,	10	00
" Steam-engine Design....................8vo,	6	00
Wilson's Steam Boilers. (Flather.).....................12mo,	2	50
Wood's Thermodynamics, Heat Motors, etc..............8vo,	4	00

TABLES, WEIGHTS, AND MEASURES.

FOR ACTUARIES, CHEMISTS, ENGINEERS, MECHANICS—METRIC TABLES, ETC.

Adriance's Laboratory Calculations......................12mo,	1	25
Allen's Tables for Iron Analysis.........................8vo,	3	00
Bixby's Graphical Computing Tables....................Sheet,		25
Compton's Logarithms................................12mo,	1	50
Crandall's Railway and Earthwork Tables................8vo,	1	50
Egleston's Weights and Measures......................18mo,		75
Fisher's Table of Cubic Yards.......................Cardboard,		25
Hudson's Excavation Tables. Vol. II...................8vo,	1	00
Johnson's Stadia and Earthwork Tables.................8vo,	1	25
Ludlow's Logarithmic and Other Tables. (Bass.).......12mo,	2	00
Thurston's Conversion Tables..........................8vo,	1	00
Totten's Metrology....................................8vo,	2	50

VENTILATION.

STEAM HEATING—HOUSE INSPECTION—MINE VENTILATION.

Baldwin's Steam Heating..............................12mo,	2	50
Beard's Ventilation of Mines..........................12mo,	2	50
Carpenter's Heating and Ventilating of Buildings..........8vo,	3	00
Gerhard's Sanitary House Inspection.............Square 16mo,	1	00
Mott's The Air We Breathe, and Ventilation............16mo,	1	00
Reid's Ventilation of American Dwellings..............12mo,	1	50
Wilson's Mine Ventilation.............................16mo,	1	25

MISCELLANEOUS PUBLICATIONS.

Alcott's Gems, Sentiment, Language............Gilt edges,	$5 00
Bailey's The New Tale of a Tub.........................8vo,	75
Ballard's Solution of the Pyramid Problem.............8vo,	1 50
Barnard's The Metrological System of the Great Pyramid..8vo,	1 50
Davis's Elements of Law..............................8vo,	2 00
Emmon's Geological Guide-book of the Rocky Mountains..8vo,	1 50
Ferrel's Treatise on the Winds........................8vo,	4 00
Haines's Addresses Delivered before the Am. Ry. Assn...12mo.	2 50
Mott's The Fallacy of the Present Theory of Sound..Sq. 16mo,	1 00
Perkins's Cornell University....................Oblong 4to,	1 50
Ricketts's History of Rensselaer Polytechnic Institute.....8vo,	3 00
Rotherham's The New Testament Critically Emphasized. 12mo,	1 50
" The Emphasized New Test. A new translation. Large 8vo,	2 00
Totten's An Important Question in Metrology............8vo,	2 50
Whitehouse's Lake Moeris............................Paper,	25
* Wiley's Yosemite, Alaska, and Yellowstone............4to,	3 00

HEBREW AND CHALDEE TEXT-BOOKS.

FOR SCHOOLS AND THEOLOGICAL SEMINARIES.

Gesenius's Hebrew and Chaldee Lexicon to Old Testament. (Tregelles.)................Small 4to, half morocco,	5 00
Green's Elementary Hebrew Grammar..................12mo,	1 25
" Grammar of the Hebrew Language (New Edition).8vo,	3 00
" Hebrew Chrestomathy..........................8vo,	2 00
Letteris's Hebrew Bible (Massoretic Notes in English). 8vo, arabesque,	2 25
Luzzato's Grammar of the Biblical Chaldaic Language and the Talmud Babli Idioms.........................12mo,	1 50

MEDICAL.

Bull's Maternal Management in Health and Disease.......12mo,	1 00
Hammarsten's Physiological Chemistry. (Mandel.)........8vo,	4 00
Mott's Composition, Digestibility, and Nutritive Value of Food. Large mounted chart,	1 25
Ruddiman's Incompatibilities in Prescriptions............8vo,	2 00
Steel's Treatise on the Diseases of the Ox.... ....8vo,	6 00
" Treatise on the Diseases of the Dog...............8vo,	3 50
Worcester's Small Hospitals—Establishment and Maintenance, including Atkinson's Suggestions for Hospital Architecture..12mo,	1 25